Known Unto Us

A Hidden History of War's Caretakers

Sara Madden

United States of America
www.saramadden.com

ISBN: 978-1-946505-14-9 (paperback)
ISBN: 978-1-946505-13-2 (hard copy)
ISBN: 978-1-946505-15-6 (eBook)

Printed in the United States of America
10 9 8 7 6 5 4 3 2 1

Cover design by Sara Madden
Cover images @ Freepik
Book design and interior layout by Sara Madden
Author photograph @ Zaugg Studios

To the silent warriors.

To the ones known by many names—caretakers, gravediggers,
stretcher-bearers, mortuary men, record keepers,
and companions of the fallen.

To those who dug with trembling hands, who carried
memory in ledgers and silence in their hearts.

And to the ones still among us—witnesses in uniform and out,
who continue the quiet labor of honoring the dead.

This book is for you.
You remembered them.
May we now remember you.

"We have come for the bodies of the slain, wishing to bury them
in observance of the universal law..."
—Euripides, *The Suppliants,* 423 BCE

"He rests in honored glory, an American soldier
known but to God."
—Inscription, Tomb of the Unknown Soldier

Contents

AUTHOR'S NOTE

I come to this work as a witness.

My background is in the American Red Cross, in International Humanitarian Law, Service to Armed Forces, and Restoring Family Links. I have sat with veterans who could not speak of what they carried. I have helped families search for the missing. I have studied the laws meant to protect the dead in war—and I have seen how often those laws fail.

This book is not about combat. It is about what comes after: the work of care, recovery, and remembrance, carried out across centuries by people whose labor has rarely been recorded and even more rarely honored. Some details in these pages were unavoidable. Many were not. Where the dead were created by deliberate cruelty or willful indifference, I have named that clearly. The caretakers live with what the perpetrators leave behind. That truth belongs in the record.

This volume is the first in a larger series examining the hidden human histories of war—not the strategies or the battles, but the hands that tended the wounded, buried the dead, carried memory, and endured long after the fighting stopped.

This is not neutrality. It is moral witness.

Prologue

A Family of Witnesses

"The dead don't stay buried. They live in the ones who carried them."

I did not set out to write this book.

It found me in Belgium, on a quiet road outside Bastogne, in the fall of 2023. My husband and I had been tracing the path of his great-uncle—a young soldier who landed in Normandy on New Year's Eve, 1944; fought in the Battle of the Bulge; and was killed in March of 1945, just weeks before the war in Europe ended. We knew he had died near Bastogne. We knew he had been buried somewhere in the area. What we didn't know, until we began searching the records, was what had happened after his burial.

He had been buried again.

He'd first been laid to rest by his comrades in the cold and muddy ground of the Ardennes in Foy, in a temporary grave marked with a wooden cross and whatever words young men could find for young men. Then, in 1949, when the cemetery closed, he had been unearthed—disinterred by other young soldiers, then processed, documented, and shipped home to Schofield City Cemetery, Utah, for permanent burial.

I stood in that field outside Foy and tried to imagine it. Boys burying boys. Then boys digging up boys. The same hands that had fired rifles and dug foxholes were now lifting the decomposed remains of friends, classmates, and brothers-in-arms. Wrapping them. Tagging them. Sending them home.

And then doing it again. And again. Grave after grave. Name after name.

My heart broke.

Not just for my husband's great-uncle, whose name I now carried with me. But for the ones who buried him. For the young soldiers tasked with this sacred, terrible labor of gathering the dead, honoring them, and then returning to a world that would never ask what that work had cost.

I realized, standing there, that I had never thought about them. In all the war films I had watched, all the history books I had read, and all the Memorial Day speeches I had heard—no one had talked about the burial crews. The stretcher-bearers. The gravediggers. The clerks who typed the names. The medics who closed the eyes.

They were invisible.

I searched for them in the historical record—in military archives, in regimental histories, and in the official accounts of every war America has fought. And I found them, but barely. They appear in footnotes. In passing mentions. In oral histories collected decades too late. The gravediggers, the mortuary clerks, the nurses who folded uniforms, the freedmen who sang spirituals over shallow trenches, and the women who stitched names into shrouds were everywhere in war and almost nowhere in our memory of it.

This invisibility didn't begin in World War II, I realized. It stretched back through every war. Through Korea and Vietnam, through the trenches of World War I, through the fields of Antietam and Gettysburg, through every conflict in which humans killed each other and then faced the question of what to do with what remained.

Someone had always done this work.

And almost no one remembered them.

I come from a family of witnesses. My great-grandfather survived the trenches of World War I. He came home blind. Whether his eyes were destroyed by gas or blast, the records are unclear. But before he lost his sight, he saw things he wrote about in his memoirs: the mud, the rats, the bodies they lived alongside, slept beside, and ate beside. The dead were not removed quickly in the trenches. They became part of the landscape. And the living learned to carry on as if this were normal.

My grandfather flew gliders in World War II. He was nineteen years old. He was captured, held as a POW, and eventually released. For most of his life, he never spoke about what he had seen. But in his final days, when the barriers that silence built began to crumble, he told me about gathering the pieces of his comrades after crashes and firefights. "You pick up what you can," he said. "You try to make them whole again. But you never forget what they looked like in pieces."

He carried that memory for seventy years.

The dead deserve to be buried with honor.

And those who bury them deserve to be welcomed home.

I am not a soldier. I have never served in combat. But I have spent years working alongside those who carry the weight of military service.

As a volunteer with the American Red Cross Service to Armed Forces program, I visit veterans in VA hospitals, where men and women are being treated for psychological wounds that war inflicted

decades ago. Once a month, we do art with them. We listen to them. We hand out stars cut from retired American flags, and I watch these veterans—tough, stoic, trained not to show weakness—weep as they hold a piece of worn fabric that somehow still carries meaning.

As an International Humanitarian Law educator with the American Red Cross, I teach young people about the rules of war—the Geneva Conventions and the protections owed to the wounded, the prisoners, and the dead. I teach them that even in war, there are limits. That humanity does not disappear when the shooting starts. That someone must always be responsible for the bodies left behind.

As a volunteer with Restoring Family Links with the American Red Cross, I help asylum-seeking women and refugees search for family members separated by conflict and displacement. The work Clara Barton began continues. The missing are still missing, and families are still searching. Someone must still hold the thread that connects the lost to the found.

This is my inheritance. Not just the stories my family carried, but the work I have chosen to continue.

Clara Barton founded the American Red Cross at age fifty-two. I was fifty-two when I realized my life wasn't ending but beginning again. I retired after thirty years as an intellectual property paralegal and started writing full time, backing that writing with my anthropology, history, and archaeology studies and degrees.

I thought: if Clara could do this work into her nineties, in a long dress with petticoats and a corset, navigating a world that did not believe women belonged in the public sphere—then I could do it now. I could write this book. I could tell these stories. I could name the ones who had been forgotten.

This book is the result.

It is not a military history in the traditional sense. It does not focus on battles won or strategies employed. It focuses on what happens after the fighting stops. On the bodies left behind. On the hands that carry them. On the systems built to process death, and the people who inhabit those systems, often at great cost to themselves.

It is a book about caretakers.

About the gravediggers and the stretcher-bearers and the mortuary specialists. About the chaplains and the medics and the clerks. About the families who waited, the volunteers who served, and the veterans who carried the dead home in their memories long after the war was over.

It is a book about the ones we forgot to remember.

I am not a detached scholar. I am a witness. And this is my testimony to the witnesses who came before. It is my attempt—however incomplete—to bring them home at last. By the end, I hope you will never watch a war movie, attend a military funeral, or pass a cemetery the same way again.

For my grandchildren: You come from people who remembered. Carry that forward. For the caretakers: You are seen now. This book is for you.

—Sara Madden

Introduction

Antietam, September 1862. The body lay in a cornfield, one hand still reaching toward something no longer there. A farmer named Samuel found him three days after the battle, when the stench had become unbearable and the crows had already begun their work.

Samuel was not a soldier. He had no shovel, only a hoe. He had no orders, only the knowledge that a man—someone's son—could not be left like this. So he dug. The September sun bore down. The flies rose and settled. And when the hole was deep enough, Samuel rolled the body in, covered the face with his own handkerchief, and pushed the dirt back over.

He did not know the man's name. He would never know it. But before he walked away, he carved into a fence post with his pocketknife: "A soldier. Buried decent."

This book is about Samuel. And the thousands like him—across centuries, across continents, across every war humanity has ever fought—who stayed behind to do what no one else would: care for the dead.

In every war, there are warriors who never lifted a weapon. Instead, they lifted the fallen. They are the gravediggers and stretcher-bearers, the mortuary clerks and field chaplains, the nurses who closed the eyes, and the mothers who never stopped searching. They are the ones who carried not rifles but body bags. Not ammunition but silence. Not glory but grief.

We honor those who die in war with monuments and parades. But the ones who return bearing the weight of broken bodies and interrupted lives are often left behind, unseen, unspoken, and unremembered.

They did not die. But something in them did.

This is their story.

Through historical record, anthropological insight, archaeological artifacts, lived testimony, and poetic interludes, this book traces humanity's oldest burden: the duty to bury the dead. From the scorched earth of ancient battlefields to the mobile mortuary labs of modern conflicts, from scaffold burials and sacred bone caves to field chaplains and Graves Registration clerks, it follows those who have stood beside death not as soldiers, but as stewards.

It reveals what war often conceals: that care is as constant as conflict. That to bury the dead is not a footnote in war but the final chapter. And that those who write that chapter, body by body, carry its cost forever.

This is not just a military history. It is a human history. Not a story of conquest but a story of *closure*—and of those who make it possible for others while being denied it themselves.

When was the last time you attended a funeral? Did you notice who prepared the body? Who dressed them? Who made them look like themselves again? Have you ever stopped to pause when a flag-draped casket has been transferred?

Probably not. We are trained not to notice. The work of deathcare is meant to be invisible—so seamless that we can pretend death itself is tidy, contained, and manageable.

But someone did that work. *Someone always does.*

This book is a call to memory. A call to reverence. A call to remember the rememberers.

Because when we remember *them*, we reclaim something war tries to take: our shared humanity.

And somewhere in Maryland in 1862, a farmer whose name was never recorded buried seventeen Union soldiers in his cornfield. He marked each grave with a stone. By spring, the stones had sunk into the mud. The names—if he ever knew them—are lost.

PART I

The Legacy of War and Burial

"Before the first weapon, there was the grave."

Before there were states, before there were formal militaries, there were battles. And after every battle, no matter how small or savage, there were the bodies.

The first act of civilization is not conquest. It is *care*.

Across cultures and across time, the burial of the dead has signaled not only the end of violence, but the reassertion of humanity. To bury is to name. To claim. To restore. It is an act of resistance against chaos, and a ritual that makes grief, somehow, bearable.

This part of the book descends into the ancient world, into the myths that gave burial its moral weight, into the battlefields upon which truces were called so enemies might gather their dead, and into the underground chambers where communities organized themselves around the sacred task of tending the fallen. We will walk the corridors of Roman catacombs, where the *fossores*, gravediggers, carved resting places into stone and prepared bodies for eternity. We will trace the emergence of burial as law, as legacy, and as labor, from the ochre-dusted graves of Shanidar Cave to the codification of war burial as a universal right.

Along the way, we meet the earliest caretakers: the women who washed the dead, the laborers who carried them, the scribes who recorded their names, the priests who blessed their passage. These were not heroes of conquest. They were stewards of memory. And their work, quiet, unglamorous, essential, laid the foundation for everything that would follow.

Here, we begin to understand what it means to bury the fallen. And more importantly, what it means to live with the memory of those we buried. Because long before dog tags or military cemeteries, before the Geneva Conventions or Graves Registration clerks, there was a hand reaching toward the body. And a voice that said: *You will not be left behind.*

Reflection

I visited Rome for the first time in 2025 while still drafting this volume. While in the city, my husband and I took a tour of 2nd-century catacombs and crypts. It was there that I saw the reverence of the care for the dead, so ancient, so neutral, so carefully tended. Across the deaths of the very young, the very old, and those who succumbed to illnesses that swept through the city, one belief remained: the importance of caring of their dead and remembrance. This was when I knew I needed to show the anthropological side of care in this book, from the beginning of our human existence. I needed to show that our first matter of urgency was always care.

Chapter One

From Myth to Mandate: Burial in Ancient Wars

"We descend not into darkness, but into memory."

Interlude: Beneath the City

The stairs spiral down into the earth, narrow, worn smooth by centuries of feet. Above, the city roars with commerce, with empire, with the business of the living. But here, beneath the streets of Rome, the air grows cool and still. The smell of damp stone replaces the dust of the Forum. Torchlight flickers against walls carved with names, dates, prayers. The dead line every passage, tucked into loculi, horizontal slots cut into the tufa, stacked three or four high, stretching into the dark like shelves in a library of souls.

This is no place of abandonment.

It is a city beneath the city. A sanctuary. A labor of love.

For three centuries, men called *fossores* worked these tunnels. They were the gravediggers of early Christian Rome, laborers who carved resting places into volcanic rock, carried bodies down the winding stairs, and sealed the tombs with tiles and prayers. Their images survive on catacomb walls: figures holding picks and lamps, depicted not with shame but with honor. They were workers, yes. But they were also witnesses. Guardians of the threshold between worlds.

Deeper still, past the burial niches, lay chambers built for care. Rooms where bodies were washed and anointed. Spaces where families gathered to share meals with the dead. *Refrigeria*, they called them, rituals of communion that blurred the boundary between presence and absence. The dead were not hidden. They were tended, fed, remembered.

The catacombs teach us that burial has always required infrastructure. That grief has always demanded community. That even in the shadow of empire, ordinary people organized themselves around one sacred task: ensuring that no one passed from this world unmarked, unnamed, or unloved.

The stairs spiral down. And we follow.

THE OLDEST RITUAL: BURYING AS A HUMAN ACT

"To Bury is to remember. And to remember is to be human."

Before war. Before religion. Before writing. There was burial.

Archaeological sites across the globe reveal that long before humans constructed cities or raised empires, they buried their dead. The oldest known intentional burial, found in Qafzeh Cave in modern-day Israel, dates back roughly 100,000 years. In it, a child was laid to rest with red ochre and tools—tokens that suggest symbolic thought and care for the dead.[1] These were not simply bodies concealed from predators or discarded to prevent disease; these were dead laid to rest with intention. With reverence. With memory.

It is the first evidence of something uniquely human: grief expressed as action.

The Neanderthals, once seen as brutish and emotionless, are now recognized among the earliest mourners in our species' story. Neanderthal remains purposefully placed in shallow graves, sometimes with the horn of an animal or the remnants of a meal, have been found at sites like La Chapelle-aux-Saints in France.[2] Some archaeologists believe flowers were used as burial offerings, early forms of what we now recognize as funeral ritual.[3] Whether or not they held religious beliefs, these early hominins exhibited something profound: a refusal to walk away from death in silence.

To bury is to resist oblivion. To place the dead in the ground is to affirm that they mattered, that they are not gone entirely if we remember them. This need becomes all the more urgent in the presence of violence.

By the time early human societies began organizing into tribes and cultures, burial had become one of the most universal human behaviors. Across Paleolithic and Mesolithic contexts, burial sites often include bodies positioned in fetal curls, facing east, or accompanied by grave goods, patterns that signal emerging spiritual frameworks. The dead were not merely discarded; they were positioned, attended to, and remembered.[4]

In ancient Jericho, burial shafts beneath domestic floors hint at a belief that ancestors remained close to the living.[5] In Catalhoyuk, a Neolithic site in modern-day Turkey, skulls were sometimes exhumed and passed around in rituals, suggesting a deep and ongoing relationship with the dead.[6] Death was not the end of presence; it was the beginning of remembrance.

This legacy is not exclusive to any one culture or people. Burial practices emerged independently across the globe. From the Pacific islands to the Andes, from the Eurasian steppes to sub-Saharan Africa, humans developed rites of burial as a means of navigating grief, asserting identity, and restoring balance, after death.[7] In many Indigenous traditions, burial was and remains a sacred act of returning the body to the earth, often tied to cosmologies in which the physical and spiritual worlds remain intertwined.

In many societies, burial was not a state function or a specialized role, but a kinship obligation carried by family, clan, or community. The dead were prepared by those who knew their names, their stories, and their place within a lineage, not by institutions or officials. Memory, rather than record, served as the primary means of preservation. This distinction matters. Before burial became assigned labor, it was intimate work, embedded in relationship rather than bureaucracy. The caretakers were not strangers. They were kin.

Burial also became political. To bury one's kin was an act of sovereignty, control over the body, the land, and the narrative of life. Denying burial, conversely, was a form of subjugation and domination. The Athenians, for instance, viewed the denial of burial not only as dishonorable, but as a violation of natural law, something Euripides dramatically portrays in *The Suppliants.*[8]

As the act of war emerged between human groups, the act of burial took on new dimensions. It became a way to reclaim the humanity that war, by design, tried to erase. In killing, men became monsters. But in burying, even the enemy, they reclaimed their humanity.

And yet, the capacity to bury did not always translate into equality in burial. Even in ancient times, not all bodies received the same care. Elites were interred with lavish goods, while commoners were often buried without fanfare. Women and slaves were frequently omitted from monumental tombs. In early Greece, burial mounds commemorated warriors; the names of their wives, nurses, and laborers rarely survived.[9] The division between who was buried and who was forgotten began early and would shape the politics of remembrance for centuries to come.

But there is another division the record obscures: not only who was *buried*, but *who* buried. Across cultures, the work of preparing the dead most often fell to women, enslaved people, or ritual outsiders, those whose labor was essential, intimate, and structurally invisible. They washed the bodies. They wrapped the shrouds. They dug the graves. And when the mourning was finished and the living moved on, they remained unnamed. Care was universal. Recognition was not.

Still, the act endured. Through fire and famine, conquest and collapse, burial remained one of the most consistent threads in human history. It served as a boundary, between life and death, between the living and the remembered. And as anthropologists argue, it is burial that defines our species more than any tool or art.[10]

Because war is not just the taking of life. It is the disruption of memory. And burial is how we fight to restore it. It is how we begin to make sense of the senseless. To hold what cannot be held.

To say, "You were here. You mattered. You are still with us, even in silence." Before there were armies, there were bodies. And before there were names carved in stone, there was a hand placing another into the earth. That, too, was an act of love.

Across the Ancient World: Death Care Beyond the Mediterranean

"Every civilization that waged war also learned to bury it."

The Roman catacombs and Greek battlefield truces offer only one thread in a global tapestry of deathcare. All over the ancient world, from the Nile to the Yellow River, from the highlands of Peru to the forests of West Africa, civilizations developed their own systems for tending the war dead. Across them all, commonalities emerged: specialized workers, sacred rituals, and the unshakeable belief that the dead must not be abandoned.

These were not borrowed traditions. They arose independently, separated by oceans and centuries, yet they converged on the essential truth that death in war demands more than grief. It demands labor. It demands infrastructure. It demands people willing to do the work that others cannot bear to witness.

The Nile's Sacred Workers: Egypt's Embalmers and the Craft of Eternity

In ancient Egypt, death was not an ending but a beginning, the threshold to eternal life. Preparing the body for that journey required an army of specialized workers whose skills were passed down through generations.[11]

The embalming process, which could take up to seventy days, was divided among distinct classes of practitioners. The *paraschistes* made the initial incision, a ritually polluting act that required its own purification rites afterward. The *taricheutes* performed the actual preservation, removing organs, treating the body with natron salts, wrapping the remains in linen.[12] Priests recited spells from the *Book of the Dead* while artisans prepared the canopic jars, sarcophagi, and grave goods that would accompany the deceased.

These were not casual laborers. They belonged to guilds centered in the *wabet*, or "place of purification," sacred workshops attached to temples where the boundary between the living and the dead could be safely crossed.[13] The embalmers understood themselves as servants of Anubis, the jackal-headed god who guided souls to the afterlife. Their work was simultaneously technical and theological, combining medical knowledge with ritual precision.

The social position of Egypt's death workers was complex and often contradictory. The *paraschistes* who made the first incision occupied the lowest rung. Their work was considered so polluting that Herodotus reports they were sometimes chased from the embalming house with stones after completing their task. Yet the *taricheutes* who performed the preservation enjoyed guild protections and hereditary status, passing their knowledge from father to son across generations. The priests who recited the sacred texts held the highest status, their spiritual authority elevating what might otherwise have been merely technical work into sacred service.

This hierarchy reveals an enduring pattern: those who touched the dead most intimately were often the most marginalized, while those who provided spiritual framing commanded respect. The hands that did the hardest work received the least recognition—a dynamic that would repeat across cultures and centuries.

For soldiers who fell in battle, particularly those who died defending Egypt or its pharaoh, proper burial was essential to cosmic order. A warrior whose body was lost or desecrated faced not merely disgrace but spiritual annihilation.[14] The embalmers who retrieved and restored these bodies were performing an act of salvation.

East Asia: The Ritual Masters of China and Japan

In ancient China, death rites were matters of state as much as family. The *Liji* (Book of Rites), compiled during the Han Dynasty, devoted entire chapters to proper mourning, specifying everything from the color of garments to the duration of wailing.[15] At the center of these elaborate ceremonies stood the *sangzhu*, the funeral master, who orchestrated every detail of the proceedings.

The *sangzhu* was responsible for preparing the body (washing, dressing, placing jade in the mouth to preserve the spirit), arranging the procession, coordinating with musicians and mourners, and ensuring that offerings were properly made to the ancestors.[16] For warriors who died in service to their lord, the *sangzhu* bore the additional responsibilities of certifying the manner of death, preparing documentation for the family, and ensuring that the grave goods reflected the soldier's rank and honor.

Confucian philosophy elevated these practices beyond mere custom. To fail in funeral rites was to fail in *xiao*, filial piety, the foundational virtue of Chinese society.[17] The care of the dead was essential to upholding moral order among the living.

The Warring States period (475–221 BCE) tested these ideals against the brutal realities of mass warfare. As seven major kingdoms battled for supremacy, casualties mounted on an unprecedented scale. The Battle of Changping in 260 BCE reportedly resulted in over 400,000 Zhao soldiers killed—a logistical catastrophe for any burial system. Contemporary accounts describe armies establishing temporary burial grounds in the field, with designated units responsible for collecting and interring the dead before advancing.

The Qin state, which would ultimately unify China, developed particularly systematic approaches to military death. Soldiers were issued identification tallies—wooden or bronze tags inscribed with name, unit, and hometown—that served as precursors to modern military identification. When a soldier fell, his tally was returned to headquarters, enabling the notification of his family and ensuring that his name would be recorded in state registers. This bureaucratic approach to death reflected broader Qin innovations in governance, transforming burial from purely familial obligation into state responsibility.

The social status of those who handled the dead in Chinese armies varied considerably. Officers responsible for burial records held positions of some prestige, their clerical work seen as essential to military administration. But the laborers who dug graves and transported bodies, often conscripted peasants or prisoners pressed into service, occupied far lower stations. Their names rarely appear in historical records, though their labor made military remembrance possible.

In Japan, Shinto beliefs held that death was a source of *kegare*, spiritual pollution that required careful management. Those who handled corpses occupied an ambiguous social position: necessary yet marginalized, essential yet feared.[18] Buddhist influence eventually softened these attitudes, and by the medieval period, warrior death had acquired its own elaborate protocols. Samurai who fell in battle were cremated with their swords, their ashes returned to family temples, their names recorded in registers that preserved their memory for generations.[19]

The *kaishaku*, the assistant who performed the decapitation stroke during ritual suicide, occupied a role of profound trust and intimacy. To serve as *kaishaku* was to bear witness to a warrior's final honor, ensuring that death came cleanly and that the body was handled with appropriate reverence.[20]

India: Fire, Water, and the Warrior's Release

In the Vedic traditions of ancient India, fire was the great purifier, the vehicle through which the soul (*atman*) was released from the body and returned to the cosmic cycle. The *antyesti*, or final sacrifice, required a specific class of priests to prepare the funeral pyre, recite the proper mantras, and guide the family through the complex sequence of rituals that followed death.[21]

For warriors of the Kshatriya caste, the martial nobility, death in battle was the highest honor. The *Mahabharata*, India's great epic of war, describes battlefield pyres lit while combat still raged, smoke rising as an offering to the gods.[22] Warriors who died facing the enemy were promised immediate ascent to the heavens, bypassing the long cycle of rebirth.

But even glorious death required care. The *antyesti* priests who tended battlefield cremations faced enormous logistical challenges: gathering fuel, preparing bodies, performing rites for dozens or hundreds of fallen warriors. Some texts describe mass cremations lasting for days, with priests working in shifts to ensure that no soldier was consigned to the flames without proper blessing.[23]

The caste system profoundly shaped who could perform death-care. Brahmin priests held exclusive authority over the sacred mantras, but the physical labor of preparing pyres and handling bodies often fell to those of lower castes. The *Doms*, traditionally associated with cremation grounds, occupied one of the lowest positions in the social hierarchy, yet performed one of the most essential services. Their presence at cremations was required, but their touch was considered polluting. They lived at the margins of society, often near burning ghats, simultaneously indispensable and untouchable.

This paradox—that those who enabled the most sacred transitions were themselves deemed impure—echoes across cultures. The hands that prepared the dead for their final journey rarely received honor for doing so.

Africa: Ancestors, Kingdoms, and the Community of the Dead

Across West Africa, death was understood not as departure but as transformation, the passage from visible to invisible membership in the community. The dead became ancestors, powerful spirits who continued to influence the living and who required ongoing attention, offerings, and respect.[24]

In the Nubian kingdoms along the Upper Nile, warrior burial traditions rivaled those of Egypt in their intricacy. The royal cemeteries at Meroe (c. 300 BCE–300 CE) contain pyramids built for kings and queens who led armies against Egyptian, Greek, and Roman forces. Excavations have revealed that elite warriors were buried with weapons, jewelry, and sometimes sacrificed servants—evidence of belief in an afterlife in which martial status would be preserved. The laborers who constructed these tombs formed hereditary guilds, their skills passed down through families over generations.

Farther south and east, the Aksumite Empire (c. 100–940 CE) developed distinct practices shaped by early Christian influence. The great stelae at Aksum marked royal burial sites, and church records suggest formalized procedures for burying soldiers who died during the empire's frequent military campaigns. Priests maintained registers of the dead, a practice that combined African traditions of ancestral remembrance with Mediterranean recordkeeping, creating some of the earliest documented military burial systems in sub-Saharan Africa.

Among the Akan peoples of present-day Ghana, elaborate funeral rites could last for days and involved the entire community. Specialized practitioners prepared the body, which might be displayed for extended periods while mourners gathered. *Okomfo* (priests) communicated with the deceased, determining whether the death was natural or caused by malevolent forces. Drummers, dancers, and singers performed rituals designed to guide the spirit safely to the ancestral realm.[25]

Additional considerations applied for warriors who died in battle,. Their blood had been shed in defense of the community, which created spiritual obligations that had to be carefully managed. Specific rites cleansed the pollution of violent death while honoring the sacrifice that had been made. The warriors' names were added to the oral histories recited at festivals, ensuring that their memory would pass to future generations.[26]

The Yoruba of present-day Nigeria developed similarly complex systems, with professional groups dedicated to preparing bodies and conducting funerals. *Egungun* societies maintained the connection between living and dead through masked dances in which ancestors were believed to temporarily return to the world of the living.[27]

The economics of deathcare in African societies often operated through communal obligation rather than individual payment. When a warrior fell, his age grade, lineage, or military unit bore collective responsibility for funeral expenses. Contributions of cloth, livestock, and labor were expected from all members, distributing the burden while reinforcing social bonds. This mutual-aid approach to deathcare would find echoes in Roman collegia, and, much later, in military death benefits and veterans' burial programs.

Northern Europe: Warriors of Bog and Barrow

The Celtic and Germanic peoples of northern Europe developed burial traditions as distinctive as their warrior cultures. Unlike Mediterranean civilizations with their temples and monuments, these societies often returned their dead to the landscape itself—to bogs, rivers, and earthen mounds that blurred the boundary between the natural and supernatural worlds.[28]

The bog bodies of Denmark, Ireland, and Britain offer haunting evidence of ritual death and deliberate preservation. Bogs' acidic, oxygen-poor environments can preserve bodies for millennia—skin, hair, and even stomach contents surviving intact. Figures like Tollund Man, discovered in a Danish peat bog with a leather noose still around his neck, reveal practices that may have combined sacrifice, execution, and sacred offering. Whether these individuals were warriors, criminals, or holy offerings remains debated, but their careful placement suggests communities that understood the bog as a threshold to other worlds.[29]

Scandinavian ship burials represent warrior deathcare at its most elaborate. At Oseberg and Gokstad in Norway, Viking-Age chieftains were interred in full-sized ships, surrounded by weapons, tools, textiles, and even sacrificed animals and servants. The ship itself served as conveyance to the afterlife, its construction and burial requiring the coordinated labor of entire communities. Craftsmen built the vessels, workers dug the burial mounds, and priests or priestesses performed the rituals that released the dead to their journey.[30]

For ordinary warriors who fell in battle, the practices were simpler but no less meaningful. Germanic peoples often cremated their dead, gathering the ashes into urns buried beneath modest mounds. The Anglo-Saxon cemetery at Sutton Hoo reveals the range of burial wealth from the magnificent ship burial of a king to the simple graves of his retainers, each positioned according to status but all within the same sacred precinct.[31]

Who performed this labor? The sagas speak of women preparing bodies for burial, washing the corpse and dressing it in finest garments. Specialized gravediggers appear in some accounts, their role suggesting that even in these societies, death work required particular skills and perhaps ritual qualification. The grave goods that accompanied warriors—swords, shields, drinking horns—had to be selected, arranged, and sometimes ritually "killed" (bent or broken) so they could accompany their owner to the next world.[32]

These northern traditions remind us that burial infrastructure need not mean stone monuments or temple workshops. A community that knew how to dig a mound, build a pyre, or weight a body for the bog possessed its own sophisticated death-care system—passed through oral tradition, enacted through collective labor, and recorded not in writing but in the landscape itself.[33]

The Americas: Warriors of the Sun and the Mountains

The great civilizations of pre-Columbian America developed some of history's most elaborate death cults, many centered specifically on warrior death.

For the Aztecs, warriors who died in battle or women who died in childbirth (understood as a form of combat) achieved the highest afterlife. They became companions of the sun, transformed into hummingbirds and butterflies who accompanied Tonatiuh across the sky. Their bodies were cremated with great ceremony, their ashes placed in special urns, their names recorded by temple scribes.[34]

The preparation of warrior bodies fell to specialized priests who understood the cosmological significance of their task. Each element—the arrangement of the pyre, the offerings placed with the body, the prayers recited over the flames—carried theological weight. To err was to risk not just personal disgrace but disruption of the cosmic order that required constant feeding through sacrifice and proper ritual.[35]

The Inca developed mummification practices that preserved the bodies of rulers and elite warriors as *mallqui*, ancestor mummies who continued to "live" among their descendants. These mummies were housed in special quarters, attended by servants, consulted on important decisions, and paraded through the streets during festivals.[36] The care of the *mallqui* required dedicated staff who bathed, dressed, and fed the preserved remains.

For ordinary warriors, burial in the Andes involved returning the body to Pachamama, Mother Earth, with appropriate offerings. Textile workers prepared the elaborate wrappings; potters created the vessels for grave goods; priests performed the rituals that ensured safe passage to the afterlife. Deathcare was a community enterprise, involving dozens of specialized hands.[37]

The Universal Pattern

What emerges from this global survey is not uniformity but resonance. Across vastly different cultures, certain patterns repeat:

Specialized death workers emerge in every complex society. They are embalmers, priests, funeral masters, gravediggers, people whose role is to stand at the threshold between living and dead.

Warrior death receives particular attention, requiring additional rites, enhanced honors, and careful management of the spiritual pollution or power that violent death creates.

Infrastructure develops to support deathcare. Sacred spaces, professional guilds, training systems, and ritual protocols emerge.

Naming and recording become essential practices, preserving the identity of the dead against the anonymity of mass death.

And across nearly every culture, a troubling pattern also surfaces: those who perform the most intimate labor of deathcare often occupy marginalized social positions. The hands that touch the dead become, in many traditions, ritually polluted or socially suspect. The essential becomes the untouchable. This paradox—that sacred work renders its workers profane—will echo through the centuries that follow, shaping how we honor (or fail to honor) those who carry the burden of war's aftermath.

These are not accidents or borrowings. They are independent discoveries of the same fundamental truth: that human communities cannot survive without systems for honoring their dead. And at the center of those systems, always, are the caretakers, the workers whose labor makes remembrance possible.

War, Honor, and the Dead in the Ancient World

"Even the enemy is sacred when dead."

In the ancient world, war and death walked hand in hand; burial walked just behind.

Among the Greeks, the burial of the war dead was not merely a formality but was a declaration of humanity amid chaos. Funeral rites, whether simple or elaborate, were seen as an obligation owed to the dead by the living, and the failure to perform them constituted a moral transgression against the cosmos itself.[38]

In the *Iliad*, Homer gives extended space not only to the deaths of heroes, but to their burials. After Achilles kills Hector, he defiles the Trojan's body by dragging it behind his chariot, an act of dishonor. But even Achilles, the wrathful son of Thetis, is eventually moved by King Priam's plea for his son's body. Priam, aged and trembling, kisses the hands that killed his son and begs not for revenge, but for the chance to bury him.[39] Achilles relents, not because of any treaty, but because some part of him recognizes the sacredness of death. It is one of the most humanizing moments in all of Greek literature.

This ethic was not confined to myth. Classical Athens maintained an official policy for recovering and honoring its war dead. After each campaign, the bodies of fallen soldiers were returned to the polis, and public funerals were held in the Kerameikos cemetery.[40] The *epitaphios logos*, a ceremonial funeral oration, was delivered before the city, reaffirming civic unity and the moral duty of remembrance.[41]

Burial was not merely personal. It was political. Elsewhere in the ancient world, burial practices following war reflected a balance between pragmatism and piety. The Assyrians, known for their brutality in battle, still observed rites for their own fallen.[42] The ancient Persians practiced sky burial, returning warriors to the elements, but also maintained necropolises, like Naqsh-e Rustam, where royal and military dead were honored with grandeur.[43]

In China during the Warring States period, philosophical texts like *The Book of Rites* emphasized the moral importance of honoring the dead, even in conflict. Armies were sometimes allowed temporary ceasefires to collect and bury their fallen.[44] This idea was codified under Confucian thought, which prioritized ritual and familial piety, extending the duty of care beyond life.[45]

Many Indigenous societies across the Americas also held deep respect for war dead. Among the Haudenosaunee (Iroquois), warriors who died in battle were buried with ceremony, and if possible, the bones of the dead were retrieved later and brought home to be reinterred in collective ossuaries.[46] Burial was a way of reclaiming the fallen and restoring balance between the living and the spirit world.

The ancient Romans institutionalized military deathcare. Legions had designated burial officers and field rituals. A proper burial, *sepultura iusta*, was essential not just for the individual, but for the army's morale.[47] Roman military tombstones are among the most detailed of antiquity, listing names, ranks, ages, and even service histories. These epitaphs reveal a bureaucratic reverence for death and a state-sponsored commitment to memory. Yet Roman honor often stopped at the empire's border, as enemy bodies were frequently left unburied as a display of dominion.[48]

Still, exceptions abounded. Numerous historical records describe moments of humanity amid horror, enemy soldiers risking their lives to return a fallen comrade to his side, or ceasefires negotiated solely for the purpose of burial. The earliest recorded battlefield truces for burial appear in the Peloponnesian War, when both Athens and Sparta agreed to temporary halts in combat to retrieve their dead.[49] These truces were not strategic; they were sacred.

In many cultures, even the enemy dead were honored. The Zulus of southern Africa, fierce warriors in the 19th century, believed in burying fallen foes to appease ancestral spirits.[50]

Among the Norse, sagas tell of enemies granting honorable cremations to worthy adversaries.[51] These acts were not weakness. They were warrior codes that extended dignity beyond the moment of death.

But as ever, the ideal and the reality did not always align.

Class, gender, and power determined who was buried and who was forgotten. Tombs were built for kings and generals. Poets praised the fallen heroes. But the laborers, camp followers, enslaved men and women, and the countless others who perished in the shadows of empire? Their names are largely lost.

The burial of the elite affirmed their legacy. The burial, or lack thereof, of the marginalized revealed the fault lines in remembrance.

Still, the through line remains: across time and place, the burial of the dead—especially the war dead—was seen as more than duty. It was a rite. A ritual of rehumanization in a world fractured by violence.

Even in ancient wars, the dead taught the living. To pause. To kneel. To remember.

These ancient customs (the battlefield truce, the sacred obligation to bury even enemies, the belief that denial of burial was a violation of divine law) would echo through centuries of conflict. They would eventually be codified into international law in the Lieber Code of 1863, the Geneva Conventions, and the Hague Accords. But that formalization came only after the horrors of industrial warfare made clear what the ancients already knew. The next chapter traces that evolution from sacred custom to legal mandate.

Cities of the Dead: The Infrastructure of Ancient Care

"They built not just graves, but systems. Not just tombs, but communities of remembrance."

To bury the dead in the ancient world was never a solitary act.

It required hands, many hands. It required space, tools, knowledge, ritual. And in the great civilizations of antiquity, it required something more: infrastructure. The elaborate burial systems of Egypt, Mesopotamia, and Rome were not accidents of culture. They were deliberate constructions, organizational responses to the most universal of human problems.

What happens when someone dies?

Who prepares the body?

Who carries it?

Who ensures the name survives?

In Rome, these questions produced answers that would echo through millennia.

The Fossores: Rome's Hidden Death Workers

The *fossores* of Rome's catacombs, whose work we glimpsed in this chapter's opening, represent one of history's earliest documented examples of specialized deathcare workers. But beyond their practical labor of excavation and burial lies a more complex story about the social position of those who tend the dead.

Unlike the *paraschistes* of Egypt, who were chased from the embalming house after their polluting incisions, the Roman *fossores* appear to have enjoyed a measure of respect within their communities. Catacomb frescoes depict them with their tools—pickaxe, lamp, measuring rod—rendered with the same artistic dignity afforded to other skilled tradesmen.

Inscriptions praise individual *fossores* as "well-deserving" or "faithful." Some were buried in the same catacomb passages they had carved, their graves marked with their names and trades.[52]

This relative status likely reflected the Christian communities' theology of death and resurrection. Where pagan Roman tradition had considered contact with corpses polluting, early Christians emphasized the body as a temple awaiting resurrection. The *fossores'* work was not contamination but preparation—readying the faithful for their eventual rising. Their labor served a sacred purpose.

Still, questions of status were never simple. Many *fossores* appear to have been freedmen or members of the lower classes, people for whom guild membership represented one of the few paths to stable income and community belonging. Their work was skilled but also dangerous—catacomb passages could collapse, lamps could fail, and the confined spaces posed constant risks. The combination of specialized skill, community recognition, and marginal social origins created an ambiguous position: respected within the faith community, yet still defined by labor that most Romans would have declined to perform.[53]

This pattern—essential workers occupying liminal social spaces—would continue throughout history. The Civil War burial crews, the Graves Registration Service men of two world wars, the mortuary affairs specialists of modern conflicts: all would inherit something of the *fossores'* ambiguous status, valued for their work yet set apart by its nature.[54]

The Collegia Funeraticia: Burial Societies of the Ancient World

Burial carried moral weight, but it also carried cost. When families could not afford to bury their dead, communities sometimes pooled resources through mutual aid or informal burial societies; when they could not, bodies were abandoned to the margins of battlefields or cities. States occasionally intervened, but such support was uneven, situational, and often withdrawn when the dead were numerous or socially expendable. Care persisted not because it was guaranteed, but because its absence was unbearable. Where obligation failed, caretakers remained, unpaid, unnamed, and necessary.[55]

This tension, between moral duty and material burden, would define the care of the war dead for millennia. And it is here, in the gap between what *should* happen and what communities could *afford* to do, that the first institutions of collective deathcare emerged.

The *fossores* did not work alone. They operated within a broader network of mutual aid organizations known as collegia *funeraticia*, burial societies that ensured even the poorest Romans could receive a dignified death.[56]

These associations functioned like early insurance programs. Members paid small monthly dues, and in return, the collegium guaranteed them a proper burial: a funeral procession, a marked grave, and the ritual meals that followed interment. For enslaved people, freedmen, and the urban poor (those who could never afford private tombs), the collegia provided a path to remembrance.

The economics of these arrangements reveal much about ancient attitudes toward deathcare. Monthly dues were typically modest—affordable even for those of limited means—because the costs were distributed across the entire membership. When a member died, the collegium drew from its collective treasury to fund the funeral. This mutual-aid model meant that deathcare was neither purely a family responsibility nor a state service, but a community enterprise sustained through ongoing contribution.

Some collegia were organized around specific trades or neighborhoods; others were attached to religious cults or ethnic communities. Military units sometimes formed their own burial associations, with soldiers contributing from their pay to ensure that comrades who fell in battle would receive proper rites. The 3rd-century Lex Collegii, a stone inscription preserving the bylaws of one such society, specifies exactly what members could expect: a funeral procession with musicians, a cremation or burial according to the deceased's wishes, and a commemorative meal for surviving members.

The existence of these societies reveals a profound truth about Roman attitudes toward death: that burial was not merely a private concern but a collective responsibility. To die without proper rites was a catastrophe, not only for the soul of the deceased, but for the community that failed them. The collegia ensured that no member, however humble, would suffer that fate.

Similar mutual aid societies existed across the ancient world. In Greece, *thiasoi* and *eranoi* pooled resources for funerals. In Egypt, guilds of embalmers passed their knowledge through generations. In Han China, community associations maintained ancestral graves for families who could not afford the upkeep alone. The details varied, but the principle that death was too important to leave to individuals remained constant.[57]

These ancient models find distant echoes in modern military

death benefits. When the United States government began assuming responsibility for burying its war dead during the Civil War, it drew on precedents established over millennia: that communities owe their fallen members dignified burial, that the costs should be collectively borne, and that someone must be designated to perform the labor of remembrance. The collegia were not direct ancestors of military graves registration—too many centuries separate them—but they demonstrate that the organizational challenges of deathcare have always prompted similar solutions.

Spaces for the Sacred: Washing, Anointing, and the Architecture of Care

The catacombs were more than burial grounds. They were functional spaces designed for the full cycle of deathcare.

Near the entrances to many catacomb complexes, archaeologists have identified chambers specifically built for the preparation of bodies. These rooms once contained basins for washing, benches for laying out the dead, and niches for storing the oils and perfumes used in anointing. The work performed here was intimate and essential, the same work that would later be done by Civil War nurses, World War I morticians, and modern-day funeral directors.

Deeper within the catacombs, larger chambers (*cubicula*) served as family tombs and gathering spaces. Here, the living came not only to inter the dead but to remember them. The practice of *refrigeria*, or commemorative meals, brought families underground to share food and wine beside the graves of their ancestors. These were not morbid occasions but celebrations of continuity, affirmations that the bond between the living and the dead remained unbroken.

Some chambers were decorated with frescoes depicting scenes of paradise, resurrection, and eternal life. Others bore simple inscriptions: "In peace." "May you live in God." "To my sweetest son." These messages were not meant for tourists or historians. They were written for the dead themselves, and to the community that would continue to visit, to maintain, to remember.[58]

Names Carved in Stone: The Preservation of Identity

Perhaps the most enduring legacy of Roman burial infrastructure is its commitment to naming. Catacomb inscriptions preserve the identities of thousands of ordinary people who would otherwise have vanished from history: children who died in infancy, enslaved women whose names their owners never recorded, laborers whose work was forgotten the moment it was finished. The *fossores* carved these names into stone, transforming anonymous bodies into individuals worthy of remembrance.[59]

This was not a bureaucratic exercise. It was a moral act.

To name the dead was to insist that they mattered, that their existence had meaning, that they belonged to a community that would not abandon them to oblivion. The same impulse that drove Roman Christians to carve names in the catacombs would later drive Civil War clerks to create burial ledgers, World War I chaplains to compile casualty lists, and modern forensic anthropologists to return identities to the unnamed dead.

The infrastructure of ancient care was built not only of stone and labor but of this fundamental belief: that every life deserves to be recorded, and that recording is itself a form of love.

The Continuity of Care

The catacombs are not a quaint relic of the past but a mirror of our present.

The *fossores* who carved graves by lamplight did the same work as the Graves Registration men who dug trenches in France. The *collegia* that pooled resources for the poor anticipated the military cemeteries that promised equal dignity to every soldier, regardless of rank. The women who washed bodies in underground chambers were ancestors to the nurses and morticians who would perform the same sacred task in field hospitals and mortuaries 2,000 years later.

The infrastructure changes. The tools evolve. But the labor remains the same.

Someone must prepare the body.

Someone must carry it.

Someone must ensure that the name survives.

And in every age, those who do this work become the invisible foundation upon which memory is built.

Spotlight: Shanidar IV: The Flower Burial

In a cave nestled in the Zagros Mountains of northern Iraq, archaeologists in the 1960s uncovered the remains of a Neanderthal male now known as Shanidar IV. What set his burial apart was the presence of pollen clusters from wildflowers, suggesting a deliberate funerary act. Some believe these blooms were placed around him as a final tribute, indicating that symbolic burial may have originated not with modern humans, but with our early cousins.[60]

Though contested by some scholars, the Shanidar Flower Burial remains one of the most poetic pieces of the archaeological record, a whisper from deep time reminding us that reverence for the dead long predates written history.It is, in many ways, the first known act of mourning.

Spotlight: The Wabet – Egypt's Houses of Purification

"They worked in the space between life and eternity."

In the shadow of Egypt's great temples stood smaller, less celebrated buildings. These were the *wabet*, or "places of purification," where the sacred work of embalming took place.

The *wabet* were not public spaces. Entry was restricted to the initiated—the *taricheutes* who performed the preservation, the *paraschistes* who made the first incision, the priests who recited the spells that would guide the soul through the underworld.[61] The work within was both technical and theological, a seventy-day process during which a corpse was transformed into an eternal vessel.

The embalmers removed the organs (storing them in canopic jars inscribed with protective deities), treated the body with natron salt, anointed the flesh with oils and resins, and wrapped the remains in hundreds of yards of linen. At each stage, specific prayers were recited. The Opening of the Mouth ceremony restored the deceased's ability to speak, eat, and breathe in the afterlife.[62]

Guild secrecy protected the embalmers' knowledge, the precise ratios of natron, the techniques for preserving facial features, and the methods for positioning the body. This expertise passed from father to son, creating dynasties of death workers whose skills were essential to Egyptian civilization.[63]

For soldiers who fell in battle, the embalmers' work was particularly urgent. A warrior's body, if not properly preserved, might trap his soul in the land of the living or condemn him to wander forever in darkness. Embalmers who retrieved fallen soldiers from distant battlefields, sometimes mummifying them in temporary field stations before transport, were performing acts of spiritual rescue.[64]

One papyrus from the New Kingdom records payment to embalmers who traveled to Nubia to recover and preserve the bodies of soldiers killed in Pharaoh's campaigns. They were gone for months. When they returned, they brought not just mummified remains but detailed records: names, ranks, the circumstances of each death.[65] They were bureaucrats of the afterlife, laboring at the edge of empire so that the dead might find their way home.

The *wabet* is gone now, its secrets largely lost. But its purpose endures in every morgue, every mortuary, every place where skilled hands prepare the dead for their final journey. The embalmers of Egypt were the first to systematize what every culture eventually learns: that the dead require care, and that care requires training, tools, and sacred space.

Artifact: The Jade Burial Suit of Prince Liu Sheng

"They sewed him in stone so his spirit would never decay."

In 1968, archaeologists excavating a Han Dynasty tomb at Mancheng, Hebei Province, discovered something unprecedented: the complete jade burial suit of Prince Liu Sheng, who died around 113 BCE.[66]

The suit consists of 2,498 pieces of polished jade, each hand-cut and drilled, joined together with nearly two and a half pounds of gold wire. It covers the prince from head to toe—face mask, torso covering, gloves for each finger, separate pieces for each foot. A second suit, belonging to his wife, Dou Wan, was found nearby, crafted from 2,160 jade pieces sewn with gold and silver thread.[67]

The labor required to create such suits was staggering. Artisans first carved each jade piece to precise specifications, then drilled the tiny holes through which the wire would pass. The pieces had to fit together seamlessly, following the contours of the body. Specialists estimate that a single suit required the equivalent of ten years of work by skilled craftsmen.[68]

But why jade? Chinese tradition held that jade (*yu*) possessed the power to preserve the body and protect the spirit. The character for jade appears in numerous words associated with purity, nobility, and immortality. To encase a body in jade was to armor it against decay, to ensure that the physical form remained intact for the soul's eternal journey.[69]

The suits were reserved for royalty and the highest nobility. Their creation required not just artisans but an entire bureaucratic apparatus: miners to extract the jade, workshops to cut and polish it, officials to record the specifications, guards to protect the precious materials. This was state-sponsored deathcare on a monumental scale.[70]

Prince Liu Sheng was a military commander who had led campaigns against the Xiongnu nomads threatening China's northern frontier. His jade suit was not just a burial garment but a statement of imperial power, the empire's pledge that those who served it would be honored in death with the most precious materials the earth could offer.

The practice of jade burial suits lasted for roughly four centuries before being banned as too extravagant. The suits that survive—fewer than twenty—and as testaments to the extraordinary lengths ancient civilizations would go to protect and preserve their honored dead.[71]

They also remind us that behind every artifact lies invisible labor: the miners, the carvers, the drillers, the assemblers, the officials, the priests, hundreds of hands working across years to prepare a single body for eternity. The prince's name survives. The names of those who sewed him in stone do not.

The First Duty After War

"Before the speeches, before the monuments—there was the grave. And someone had to dig it."

War does not end with victory. It ends with burial. Not when the last gun is fired or the last treaty signed, but when someone bends their back to the earth and makes space for the dead. When the body is retrieved. The name written. The grave marked. This is the first duty after war—not reconstruction, not reconciliation, but remembrance. Across millennia, across cultures, this duty has remained constant. Someone has always stayed behind. Someone has gathered the broken. Cleaned what could be cleaned. Laid the body down. Marked its passing with stone, soil, or prayer. Sometimes that someone was a fellow soldier. Sometimes a mother, a priest, a stranger, a gravedigger with no name. Sometimes the burial was swift and crude. Sometimes it was tender. Painstaking. Holy.

But it was done. Because it had to be. And here is what we rarely ask: *who were they?* Not the fallen—we build monuments to the fallen. We carve their names in marble. We drape their coffins in flags. But the ones who dug? The ones who washed the blood from broken limbs? The ones who whispered prayers over bodies too shattered to identify?

We do not know their names. We never thought to ask. This book is an asking.

The Unnamed: Somewhere outside Troy, circa 1180 BCE, a woman whose name Homer never recorded walked among the dead after the Greeks had sailed. She was not a queen. She was not mourned in verse. But she closed the eyes of boys who had no mothers left to weep for them. She buried what the victors left behind.

She is one of the first caretakers we do not know. She will not be the last.

Reflection

The more I learn about the rules of war, the more I realize how much they matter—and how often they are broken. As an International Humanitarian Law instructor with the American Red Cross, I teach young people about the Geneva Conventions. I explain that even in war, there are limits. That the wounded must be cared for. That prisoners must be protected. That the dead must be treated with dignity—identified, recorded, and returned to their families when possible. These are not suggestions. They are laws. And they exist because previous generations understood what happens when war has no boundaries: atrocity, erasure, and the complete destruction of human dignity. Teaching these principles to young people—the world's future humanitarians—has become central to my volunteer work. I watch their faces as they learn that the rules of war protect everyone, including enemies. I see them grapple with the idea that humanity persists even in conflict, that someone must always be responsible for the bodies left behind. Today, as I write this, these rules are being violated around the world. Bodies left unburied in conflicts that generate headlines for a week and then fade from public attention. The laws exist. The question is whether we have the will to enforce them and whether we are raising a generation that understands why they matter. This next chapter traces how those laws came to be. But for me, the history is never abstract. It is alive in every classroom where I teach, every young person who asks, "Why does this matter?" and every moment when the rules of war stand between humanity and barbarism. The answer is always the same: because the dead deserve dignity. And because the living must insist upon it.

Chapter Two

The Laws of War: Collecting the Dead

Part A: The Roots-Ancient Customs Through the Medieval Era

"In myth, burial was a sacred obligation. In modern war, it became a legal one."

From Sacred Custom to Legal Code

Long before treaties or tribunals, humanity had laws of the dead. Not written on parchment or enforced by governments, but honored in gesture, innately brought to consciousness, in grief, in rituals passed down through generations.

To bury the fallen was not merely kindness, it was obligation. Not to the state, but to the soul.

Ancient civilizations enshrined this duty in story and custom. In *Antigone,* Sophocles dramatizes the moral agony of a sister forbidden to bury her brother by royal decree. "I was not born to hate together,"

Antigone tells Creon. "But to love together."[1] Her decision to bury the body, defying state law, affirmed a higher law older than any crown or code. A law written in the process of mourning.

But as warfare scaled, as armies grew, and as the dead numbered in hundreds or thousands, sacred custom struggled to keep pace. The intimacy of war, the knowing of the dead, gave way to anonymity. Industrialized killing demanded something new: a system. A structure. A codification of memory.

This shift began slowly.

During the medieval era, the chivalric codes that governed knightly behavior introduced the expectation that even enemies deserved respect in death, at least if they were of noble blood.[2] Burial was considered a mark of civility and shared martial identity. Yet this ethic rarely extended to the common foot soldier, let alone to civilians or conquered peoples. Class, rank, and race still determined whether one was buried or abandoned to the carrion.

Still, the idea that there was honor in honoring the dead, even in war, endured.

By the 17th and 18th centuries, Enlightenment thinkers began to articulate war as a legal and moral condition rather than sheer chaos. Jurists like Hugo Grotius, the so-called "father of international law," argued that even war must have boundaries, that human dignity could not be entirely suspended by conflict. Grotius insisted that enemy dead were owed burial, not as an act of mercy, but of justice. His ideas laid the intellectual groundwork for future codes.[3]

But it was not until the 19th century that burial became part of formal military regulation—and not until the 20th that it became enshrined in international law. The story of that transformation spans two continents, two world wars, and countless acts of both horror and conscience. It begins with a grieving father in America and a Swiss businessman in Italy.

We bury, therefore we are not beasts.

Interlude: Before the Protocols

Before the rules, there was reverence.Before the treaty, the tear.Before the commission, a hand laid gently on a shoulder.

Long before lawyers drafted codes of conduct, the dying were laid in the dirt by those who wept. They didn't ask for rank or allegiance. They just buried.

And yet their acts, the quiet, undocumented rituals, built the case for every future law. Because before the law codified what *should* be done, humans were already doing it. Not because they had to but because they couldn't bear not to.

The Chivalric Seed: Burial as a Mark of Honor

"How we treat the dead—even the enemy's dead—reveals our civilization."

In the medieval world, death on the battlefield was often a bloody, brutal, and deeply symbolic spectacle. But within the chaos of combat, a paradox existed: even in a time of relentless violence, there were rules. Among knights and nobles, these rules formed the ethical spine of chivalry, a code that exalted bravery, loyalty, and, perhaps most curiously, honor in death.

Within this framework, burial was not a universal right; it was a privilege, extended primarily to those deemed worthy by class or allegiance. Yet in that exclusivity, the chivalric tradition planted the seeds of a broader principle: that even amid enmity, death demanded deference and respect.[4]

The burial of enemies, particularly of noble rank, was not uncommon. During the First Crusade, when European knights encountered Muslim adversaries they were taught to fear and despise, there are recorded instances of battlefield truces to allow for burial.[5] These acts were often politically and spiritually motivated, but they signaled a shared recognition of human dignity beyond the sword.

In medieval France and England, fallen knights were frequently interred with ceremony, their bodies returned to families or buried on the field under cairns or stone markers.[6] A warrior's burial was both a religious rite and a public symbol, proof that the dead had fought honorably and died well. The battlefield tombstone became not only a resting place but a narrative tool for proclaiming lineage, courage, and sacrifice.

The chivalric ideal, immortalized in literature by Chretien de Troyes and later by Malory's *Le Morte d'Arthur*, emphasized that to die in combat was noble, but to be denied burial was shame.[7] Knights who abandoned fallen comrades or desecrated enemy bodies risked being branded dishonorable. In this way, burial functioned as a moral ledger: a final accounting of character.

Yet this code was far from egalitarian.

Peasants, conscripts, and enemy foot soldiers rarely received the same consideration. Their bodies were often left on the field, scavenged by animals, or dumped in mass graves. During the Hundred Years' War, chroniclers like Jean Froissart lavishly described the funerals of noblemen while glossing over the fate of thousands of common dead.[8] The hierarchy of memory was written in both armor and stone.

Still, chivalric ethic reached beyond the confines of class in some moments. During the Battle of Crecy (1346), Edward III of England ordered that French knights, though enemies, were to be buried with proper rites.[9] Not out of sentimentality, but as an overt statement of English virtue and royal dignity. The act of burial became political theater, an assertion that English arms were not only strong but just.

Chivalry, in this sense, laid a foundation. It introduced the idea, however inconsistently applied, that dignity in death must not be reserved only for the victor. That burial could be an act of power, yes, but also of principle.

The rise of Christian theology further cemented the spiritual weight of burial. The body, as a vessel awaiting resurrection, had to be treated with care.[10] Desecrating corpses—even those of heretics or heathens—was increasingly seen as an affront to God. This theological shift provided moral backing to chivalric customs, transforming battlefield burial from tradition into obligation.[11]

The legacy of chivalry in burial is not its perfection, but its principle. It introduced the idea that how one died, and how one was buried, mattered. It began to tether military might to moral responsibility. It whispered the future promise of universal rights, even if those rights were centuries away from recognition.

And though the sword often ruled, there were still moments when hands laid it down—not to fight, but to bury.

Spiritual Laws and Moral Weight

"Long before the law said to bury, the soul already knew."

Across cultures and centuries, before any formal treaties or conventions, there were laws not written in ink but in spirit. Burial was not just custom; it was cosmology. It was not enforced by generals but by gods, ancestors, and the deeply held belief that how we treat the dead determined who we were among the living. In nearly every major world religion, the burial of the dead is considered a sacred act—an essential gesture that upholds the dignity of the deceased and preserves the moral balance of the universe.

In Christianity, burial is intimately tied to the theology of resurrection. The body, though perishable, is seen as a vessel awaiting restoration at the end of days. From early martyrs in catacombs to modern-day military chaplains on the battlefield, the Christian tradition has held that to bury the dead is an act of love, service, and obedience to divine law. The corporal works of mercy, as listed in Catholic doctrine, explicitly name "burying the dead" as a holy duty.[12]

In Judaism, the mitzvah of *kavod ha-met* (honoring the dead) is one of the most sacred obligations. The ritual purification (*taharah*), prompt burial, and mourning customs that follow are not merely traditions but moral commandments. Jewish law prohibits delaying burial and insists on the equal dignity of all the dead, regardless of status. Even the enemy's body, according to Talmudic guidance, must be treated with respect.[13]

Islam, too, enshrines burial as both a religious and communal duty. The body is to be washed, wrapped in white cloth (*kafan*), and laid in the ground with the head facing Mecca.[14] Burial should occur as soon as possible after death, affirming the Islamic view that the body must be returned to the earth in peace and humility. The Prophet Muhammad himself emphasized proper treatment of the enemy dead during battle, instructing followers not to mutilate corpses and to grant burial even to those slain in opposition.[15]

In Hinduism, while cremation is the norm, the treatment of the body remains a sacred responsibility. The dead are honored through last rites (*antyesti*), which guide the soul (*atman*) on its next journey.[16] Failing to perform these rites is seen as a rupture in *dharma*, the moral order. In times of war, even the enemy was sometimes granted ritual fire to prevent spiritual unrest.

Among Indigenous cultures, burial and post-death rituals are often deeply tied to cosmology and ecological balance. The Lakota, for instance, believe that the spirit (*nagila*) must be released through sacred rites that include songs, prayers, and offerings.[17] To leave a body unburied disrupts the journey of the soul and the harmony of the land. In Andean traditions, the dead are part of a reciprocal relationship with the living—feeding the soil, guiding the harvest, and anchoring ancestral memory.[18]

These spiritual laws operate independent of national borders. They endure even when empires fall or treaties are broken. And often, it is the clergy, healers, and local individuals who ensure that these laws are followed when the state cannot or will not.

During World War II, it was not uncommon for field chaplains, monks, or even local villagers to conduct burial rites when official services were absent. In the Pacific Theater, Japanese and American soldiers alike were buried by local elders using whatever ritual elements they could muster. In Europe, Catholic priests often moved through shelled-out towns, blessing bodies with water scavenged from ruined wells.[19] These were acts of spiritual triage—improvised but deeply sincere.

Such acts reflect something enduring: a refusal to let the soul be forgotten, even if the name is lost.

And yet, spiritual laws often fill the spaces where legal ones fall short. The Geneva Conventions may outline burial protocols, but they cannot reach every battlefield, every forest, every unnamed trench. Faith, on the other hand, is carried in pockets, whispered in last breaths, and shared through gestures that do not require translation.

There is moral weight in burial that cannot be legislated. It is found in a whispered prayer over a nameless body, in a cloth draped with care, in the hand placed gently on a shoulder as someone digs the final inch of grave. To bury, in the spiritual sense, is not just to dispose of a body. It is to heal a tear in the moral fabric of the world.

Even in war—perhaps especially in war—this healing matters. Because when the world descends into chaos, burial becomes an act of faith that something sacred and enduring still exists. That death is not the end of reverence. That love can still take form, even in a shovel, a song, or a final quiet moment beside the fallen.

Part B: The Architecture of International Law-From the American Civil War to the Geneva Protocols (1863–1977)

Spotlight: Franz Lieber and the Code That Brought Law to the Battlefield

In 1863, amid the horrors of the American Civil War, the Union Army adopted General Orders No. 100, better known as the *Lieber Code*. Drafted by German-American legal scholar Franz Lieber, who had lost a son to the war, the code marked the first formal attempt by any nation to regulate conduct during armed conflict.

Lieber was uniquely positioned to understand both the theory and agony of war. Born in Berlin in 1798, he had fought against

Napoleon as a young man and carried a wound from Waterloo for the rest of his life. He immigrated to America, became a professor of political science, and raised three sons—two of whom fought for the Union, one for the Confederacy. When his son, Oscar, was killed at the Battle of Williamsburg, Lieber knew intimately what it meant to lose someone to war. His code would carry that grief into law.

The code made explicit what had previously been implied: even enemy dead must be protected. Article 75 declared that "Honorable burial should, if possible, be given to the dead of the enemy."[20] This single line transformed a moral obligation into a military duty.

The Lieber Code was revolutionary. It transformed the care of the dead from a discretionary kindness into a military expectation. Burial was no longer just ritual, it was regulation. The Code also introduced the idea of posthumous rights. A soldier, even in death, retained identity. Their body could not be mutilated or left unburied without violating not only the law of conscience, but the laws of war.

The moral clarity of Lieber's contribution made it unique. Even in war, he insisted, there must be dignity. His words reminded a fractured nation that remembrance was not weakness but the highest form of strength.

The Lieber Code became the model for later conventions—including the Geneva and Hague Conventions—by framing burial not just as mercy, but as law. It was a watershed; but it was also an American document, born of American crisis. For burial rights to become truly universal, the principle would need to cross oceans and bind nations that shared no common soil, no common flag, no common dead.

That transformation began not in a council chamber, but on a battlefield—where a Swiss businessman stumbled into a scene of suffering that would haunt him for the rest of his life and change the world forever.

The Birth of Modern Humanitarian Law

"A man lay dying in the Italian sun. And from his suffering, a movement was born."

The transformation of battlefield burial from sacred custom to legal obligation did not happen in council chambers or through philosophical debate. It began in blood, on a hillside in northern Italy, where the collision of three armies produced a horror so vast that it awakened the conscience of an entire continent.

On June 24, 1859, the forces of France and Sardinia clashed with the Austrian Empire near the town of Solferino in Lombardy. By nightfall, nearly 40,000 men lay dead or wounded across the rolling terrain. Vineyards and wheat fields had been transformed into a landscape of agony. The battle was decisive for Italian unification, but its true legacy would emerge from the eyes of a witness rather than military strategy.

Henri Dunant, a Swiss businessman, had traveled to Italy hoping to meet Napoleon III regarding a business venture in Algeria. Instead, he arrived at Solferino the day after battle and found himself surrounded by dying men—French, Austrian, and Italian alike—crying out for water, for aid, for the mercy of death.[21] The medical services of all three armies had been overwhelmed. Surgeons operated without anesthesia. Wounds festered in the summer heat. Thousands of men who might have survived with basic care instead perished from neglect. Dunant organized local women to tend the wounded, insisting on a principle that would reshape international law: *"Tutti fratelli"*—all are brothers.[22] Dunant made no distinction between friend and enemy.

A wounded Austrian received the same cup of water as a wounded Frenchman. In this simple act of impartial care lay the seed of modern humanitarian law.

Yet the principle was enacted by hands that history rarely names. The women of Castiglione delle Stiviere—farmers' wives, shopkeepers, daughters of the village—became the first responders. They tore linens into bandages. They carried water through streets choked with the dying. They held the hands of men whose languages they did not speak and whose uniforms meant nothing beside their wounds. Dunant gave voice to a principle, and these women gave it form. But when the memory of Solferino was written into treaty, their labor was absorbed into his legacy, essential, invisible, unnamed.

Three years later, Dunant published *A Memory of Solferino* (1862), a harrowing account of what he had witnessed and a proposal for organized relief. He called for the creation of voluntary relief societies in every country—trained civilians who could assist military medical services during wartime. He further proposed that these volunteers, and the wounded they served, be granted protected status under international agreement.[23]

The book electrified Europe. Military officers, physicians, philanthropists, and heads of state read Dunant's testimony and recognized that modern warfare had outpaced the capacity for human care. Something had to change.

In 1863, a committee of five Geneva citizens—including Dunant and the jurist Gustave Moynier—founded what would become the International Committee of the Red Cross (ICRC).[24] Their goal was modest but revolutionary: to establish neutral humanitarian organizations that could operate on battlefields without being targets of war.

The following year, on August 22, 1864, representatives from twelve European states gathered in Geneva to sign the first Geneva Convention, formally titled the "Convention for the Amelioration of the Condition of the Wounded in Armies in the Field."[25] Though only ten articles in length, its implications were profound.

The Convention established that military medical facilities and personnel were to be considered neutral, protected from attack, and wounded soldiers, regardless of nationality, were entitled to care. A distinctive emblem—the red cross on a white field—would mark protected persons and places.[26]

While the 1864 Convention focused primarily on the wounded rather than the dead, it established a principle that would expand in subsequent decades: that certain categories of persons and certain humanitarian acts stood outside the logic of warfare. If the wounded could be protected by international agreement, so too could the dead.

The Convention also introduced something unprecedented in the history of warfare: *accountability*. States that ratified the agreement bound themselves to specific obligations and became subject to the scrutiny of the international community. War, that most anarchic of human endeavors, had been touched—however lightly—by law.[27]

Though Henri Dunant would live to see his vision transformed into a global movement, his personal fortunes collapsed and he spent years in poverty and obscurity. In 1901, he was awarded the first Nobel Peace Prize, sharing the honor with Frédéric Passy. By then, the Red Cross had spread to dozens of nations, and the principles of Geneva had been tested in conflicts from the Franco-Prussian War to the Spanish–American War.[28]

Yet Dunant's original vision remained incomplete. The 1864 Convention said nothing directly about the dead—about their identification, burial, or return to families. These provisions would come later, built upon the foundation that Solferino's witness had laid. What began as one man's horror at preventable death would grow into a comprehensive legal framework for the care of all who fall in war, living and dead alike.

The road from Solferino to the modern Geneva Conventions would span nearly a century and require the devastation of two world wars. But it began with a single principle, spoken in a makeshift hospital among the dying: *Tutti fratelli.*

We are all brothers.

Even in war. Even in death.

THE HAGUE CONVENTIONS: CIVILIZING THE CONDUCT OF WAR

"If war could not be abolished, perhaps it could be tamed."

The Geneva Convention of 1864 had opened a door. But the room beyond remained vast and largely unregulated. While the wounded now enjoyed at least nominal protection, the broader conduct of warfare—the weapons used, the treatment of prisoners, the fate of the dead—remained governed primarily by custom, honor, and the discretion of commanders.

In the final years of the 19th century, an unlikely figure would attempt to change that. Tsar Nicholas II of Russia, motivated by a complex mixture of humanitarian concern and strategic calculation, issued an extraordinary invitation in 1898: he called upon the nations of the world to gather and discuss the limitation of armaments and the codification of the laws of war.[29]

The result was the First Hague Peace Conference of 1899, convened in the Netherlands with representatives from twenty-six states.

Though the conference failed in its primary aim of arms limitation, it succeeded in producing three conventions and three declarations that would shape international humanitarian law for generations.[30]

The most significant of these was the "Convention with Respect to the Laws and Customs of War on Land," which included the Hague Regulations—a detailed code governing the conduct of belligerents. For the first time, an international agreement addressed not merely the protection of specific categories of persons, but the general principles by which war itself should be waged.[31]

Among its provisions, the 1899 Hague Regulations established the definition of lawful combatants entitled to prisoner-of-war status; prohibitions on certain weapons deemed to cause unnecessary suffering; rules governing the treatment of occupied populations; and requirements for the humane treatment of prisoners of war.[32] Crucially, the regulations also touched upon the dead. Article 14 required that belligerents establish an information bureau to collect and transmit information about the wounded, sick, and dead who fell into their hands.[33] This seemingly administrative provision represented a profound shift. The dead were no longer merely bodies to be disposed of, but persons whose identities and fates were a matter of international concern.

The 1899 Conference also produced a Convention adapting the principles of Geneva to maritime warfare, extending protections to the wounded, sick, and shipwrecked.[34] The ocean, which had long swallowed the dead without record or ceremony, was now subject—at least in principle—to humanitarian law.

Eight years later, in 1907, a Second Hague Peace Conference convened with the broader participation of forty-four states, including nations from Asia and Latin America. This conference revised and expanded the 1899 agreements, producing thirteen conventions that addressed everything from the opening of hostilities to the rights of neutral powers.[35] The revised Hague Regulations of 1907 strengthened provisions regarding the dead. Article 19 required belligerents to issue burial certificates or authenticated lists of the dead and to ensure that burials were "honorable" and that graves were "respected" and properly maintained.[36] The regulations further required that graves be marked so they could always be found, a provision that would take on terrible urgency in the trenches of the coming world war.

The Hague framework also introduced what became known as the "Martens Clause," named for the Russian diplomat Fyodor Martens. This clause, inserted into the preamble of the 1899 Convention, declared that in cases not covered by specific regulations, civilians and combatants remained under the protection of "the principles of the law of nations, as they result from the usages established among civilized peoples, from the laws of humanity, and the dictates of the public conscience."[37]

The Martens Clause was more than legal boilerplate. It established that international humanitarian law was not merely a collection of specific rules, but an expression of fundamental principles that could be invoked even when treaties fell silent. The dead, even when not explicitly mentioned, remained under the protection of humanity's conscience.

Yet the Hague Conventions, for all their ambition, reflected the limitations of their era. They were drafted by European powers at the height of empire, and their protections often applied most robustly to "civilized" nations waging war against one another. Colonial conflicts, wars of conquest against Indigenous peoples, and civil wars often fell outside their scope.[38] The framework assumed a world of sovereign states, uniformed armies, and declared wars—assumptions that subsequent history would repeatedly challenge.

The conventions also lacked meaningful enforcement mechanisms. States pledged to honor the laws of war, but violations were addressed primarily through diplomatic protest or, in extreme cases, reprisal.

There was no international tribunal, no systematic monitoring, no accountability beyond the conscience of belligerents themselves.[39]

These limitations would become painfully evident in August 1914, when the armies of Europe marched into a war that would overwhelm every humanitarian system devised to contain it. The Hague Regulations required honorable burials and marked graves. They said nothing about what to do when bodies were buried by shellfire, obliterated by artillery, or lost in mud that stretched for miles.

And yet, the Hague framework survived. Its principles were invoked throughout the First World War, even as they were routinely violated. More importantly, its failures became the foundation for reform. The horrors of 1914–1918 demonstrated not that international humanitarian law was futile, but that it was insufficient—and that insufficiency demanded remedy.

But the essential insight of The Hague remained: that war, though it could not be prevented, need not be lawless. That even amid destruction, certain principles endured. That the dead, no less than the living, deserved the protection of international conscience.

The laws were imperfect. The enforcement was weak. But the aspiration was clear.

To be civilized, even in war.

The Truce Between Trenches: Humanity in the Chaos

"Even when nations warred, soldiers sometimes remembered they were still men."

In the brutal machinery of modern warfare, where death is calculated in tonnage and trenches, something quietly defiant persisted: the instinct to stop fighting long enough to care for the dead.

During World War I, amid the deafening chaos of industrialized slaughter, soldiers on both sides occasionally enacted what commanders and politicians could not—a truce, not for peace, but for burial. These unofficial ceasefires, spontaneous, fragile, and profoundly human, are among the most remarkable moments of the Great War. They remind us that even amid mass mechanized destruction, the soul still seeks to kneel before the fallen.

One of the earliest and most documented examples occurred during the Christmas truce of 1914, when British and German troops along the Western Front laid down their weapons, emerged from their trenches, and met in No Man's Land. They exchanged cigarettes, sang carols, and even played football. But they also buried their dead. Together. Side by side.[40]

Letters home from soldiers recount the surreal beauty of it: enemies carrying stretchers, digging graves, and sharing silent gestures of respect over the bodies of their comrades.[41] This was not a sanctioned pause. It was a rebellion of memory against the tide of annihilation. Though the Christmas truce is the most famous, such moments occurred throughout the war.

During lulls in shellfire or after particularly brutal assaults, opposing units raised white flags, sent runners, or shouted across trenches to negotiate short reprieves for body collection.[42]

These truces were dangerous—technically illegal and often opposed by high command. But soldiers, grounded in the mud and blood of war, understood something that generals sometimes forgot: death demanded dignity.

At Verdun, Ypres, and the Somme, battlefield truces allowed men to retrieve the broken bodies of friends and foes alike. In some cases, enemy soldiers carried the other's dead back across lines. In others, graves were dug in the middle of No Man's Land, uniting the dead from both sides in a final act of parity.[43]

This behavior was not new. It echoed earlier battlefield customs, such as those seen in the Peloponnesian War, when Athenians and Spartans paused to collect their dead. But in the trench warfare of WWI, where proximity, stagnation, and shared suffering defined a soldier's life, these truces became uniquely intimate.

They also became a kind of resistance.

By stepping into No Man's Land not to kill, but to bury, soldiers reclaimed their humanity—if only for a moment. They made space for grief. For reverence. For something other than survival. And in doing so, they subverted the logic of war itself.

The military bureaucracy often ignored or suppressed these stories, worried that compassion might weaken morale. But the letters survive. So do the photographs. And so does the memory of soldiers who crossed the very fields they were told to hold, not with rifles, but with stretchers.

Some historians argue that these moments were anomalies, brief flickers of sentiment in an otherwise relentless war.[44] But they were more than anomalies. They are evidence. Evidence that, even in the most dehumanizing conditions, ritual and respect can endure without compulsion or perhaps recognition.

To bury, even the enemy, was not cowardice. It was courage.

And in a war that redefined what it meant to be human, those who chose to remember—even for a day—offered a different kind of victory.

These fragile moments of humanity amid horror could not be legislated. They arose from something deeper than statute, from the same impulse that had moved Priam to beg Achilles for Hector's body, the same reverence that had shaped burial customs since Shanidar Cave. But they also revealed the inadequacy of existing law. If the Hague Conventions could not prevent the desecration of the dead on an industrial scale, something more was needed. The postwar world would try to build it.

The Geneva Conventions Reforged: From World War to Universal Law

"Each revision was written in the ink of the previous war's failures."

The Hague framework had established that war should have rules. The First World War demonstrated how easily those rules could be shattered—and how desperately they needed to be rebuilt.

Between 1914 and 1918, the nations that had gathered in The Hague to civilize warfare instead unleashed industrialized slaughter on an unprecedented scale. Poison gas drifted across trenches. Machine guns cut down entire generations. Artillery barrages buried the dead beneath the same mud that entombed the living. The information bureaus required by the Hague Regulations were overwhelmed within weeks; the requirement for marked and respected graves became a cruel impossibility when bodies disappeared into shell craters or were simply never recovered.[45]

Yet even amid this catastrophe, the framework held—barely. The International Committee of the Red Cross, operating from neutral Switzerland, managed to transmit millions of messages between prisoners and their families. The ICRC's Central Prisoners of War Agency processed information on captured soldiers and facilitated the exchange of wounded prisoners. Delegates visited camps on both sides, documenting conditions and advocating for humane treatment.[46]

When the guns finally fell silent, the international community faced a reckoning. The existing conventions had proven insufficient. Millions of families still did not know where their sons, husbands, and fathers lay buried, or whether they had been buried at all. The gap between legal aspiration and wartime reality demanded urgent remedy.

The Geneva Convention of 1929

In July 1929, representatives from forty-seven states gathered in Geneva to revise and expand the existing humanitarian framework. The conference produced two conventions: one updating the 1906 Convention on the wounded and sick, and another—entirely new—addressing the treatment of prisoners of war.[47]

The 1929 Convention relative to the Treatment of Prisoners of War was the first comprehensive international agreement devoted exclusively to POWs. Its ninety-seven articles established detailed standards for capture, internment, labor, discipline, and release. For families desperate for news, it required the systematic collection and transmission of information about prisoners—their names, their locations, their health.[48]

Crucially, the 1929 POW Convention also addressed what happened when prisoners died in captivity. Article 76 required that prisoners who died be "honourably buried" and that their graves be "respected and properly maintained."[49] Detaining powers were required to transmit death certificates and authenticated lists of the dead to the opposing belligerent. Article 77 mandated the establishment of an official graves registration service to record the location and identification of each burial.[50]

These provisions represented a significant advance. The dead were no longer merely mentioned in passing; they were the subject of specific, detailed obligations. The 1929 Convention recognized what bereaved families had always known: that knowing where a loved one lay buried was not a bureaucratic nicety but a profound human need.

The convention also formalized the role of the ICRC as a neutral intermediary, authorizing its delegates to visit camps and to receive and transmit information about prisoners and the dead.[51] The Central Prisoners of War Agency, which had operated on an ad hoc basis during the First World War, now had explicit legal standing.

Yet even as the 1929 Conventions were being ratified, their limitations were becoming apparent. They applied only to international armed conflicts between states that had ratified them. They said nothing about civil wars, colonial conflicts, or wars of national liberation. And they assumed a world of uniformed armies, clear front lines, and identifiable combatants—assumptions that the coming decades would shatter.

THE CRUCIBLE OF THE SECOND WORLD WAR

The Second World War tested the 1929 framework to destruction—and beyond.

On the Western Front, the Geneva Conventions functioned imperfectly but recognizably. American, British, and German prisoners generally received treatment consistent with the 1929 standards, though violations occurred on all sides.

The ICRC transmitted millions of messages, visited thousands of camps, and maintained records that would prove invaluable for postwar recovery efforts.[52]

But elsewhere, the conventions might as well have not existed. Japan, which had signed but never ratified the 1929 POW Convention, subjected Allied prisoners to systematic brutality—forced labor, starvation, medical experiments, and mass executions. The Soviet Union, which had also not ratified the convention, and Nazi Germany engaged in mutual atrocities on the Eastern Front that defied all humanitarian norms. Millions of Soviet prisoners died in German captivity; German prisoners in Soviet hands fared little better.[53]

And then there were the categories of persons whom the conventions had never contemplated: the civilian populations subjected to aerial bombardment, the millions murdered in concentration camps and killing fields, the partisans and resistance fighters who fought without uniforms or insignia. The Holocaust alone rendered the existing framework grotesquely inadequate. What convention could address the systematic murder of six million Jews and millions of others deemed unworthy of life?[54]

The war also exposed the inadequacy of provisions regarding the dead. Cremation in concentration camps left no graves to mark. Mass executions produced mass graves that occupying forces deliberately concealed. Bombing reduced bodies to ash. And even where traditional warfare prevailed, the scale of death overwhelmed identification and burial systems. The graves registration services of the major powers performed heroically, but millions of dead remained unidentified or unrecovered.[55]

By 1945, it was clear that international humanitarian law required not merely revision but transformation.

The Geneva Conventions of 1949

The diplomatic conference that convened in Geneva in April 1949 was the largest and most ambitious gathering in the history of humanitarian law. Representatives from sixty-four states—far more than had participated in any previous conference—met to draft what would become the four Geneva Conventions that remain in force today.[56]

The 1949 Conventions represented both consolidation and expansion. The First Convention updated protections for the wounded and sick in armed forces in the field. The Second Convention addressed the wounded, sick, and shipwrecked at sea. The Third Convention comprehensively revised the law on prisoners of war. And the Fourth Convention broke entirely new ground by establishing protections for civilians in time of war.[57]

For the care of the dead, the 1949 Conventions represented a quantum leap.

The First Convention devoted an entire article—Article 17—to the deceased. It required that parties to a conflict ensure that the dead were honorably interred, "if possible according to the rites of the religion to which they belonged." Graves were to be respected, properly maintained, and marked so they could always be found. The convention required the establishment of a Graves Registration Service to record all available information about the dead: their identity, location of burial, and cause of death. It further mandated the transmission of death certificates and lists of the dead to the opposing party, either directly or through the ICRC.[58]

Article 17 also addressed the increasingly common situation in which bodies could not be immediately buried. It required that, "as far as military considerations allow," remains be examined—preferably by a medical officer—before burial or cremation, to confirm death and facilitate identification. The dead were to be identified by their identity discs; one disc was to remain with the body while the other was forwarded to the information bureau.[59]

The Third Convention, on prisoners of war, similarly strengthened provisions for the dead. Article 120 required that deceased prisoners be honorably buried, "if possible according to the rites of the religion to which the deceased belonged," in individual graves unless "unavoidable circumstances" required collective graves. The article specified that bodies could be cremated only for "imperative reasons of hygiene, on account of the religion of the deceased or in accordance with his express wish," a provision clearly influenced by the Nazi regime's use of cremation to destroy evidence of atrocity.[60]

The Fourth Convention extended parallel protections to deceased civilians, requiring respectful treatment, proper burial, and the transmission of information to family members and national authorities.[61]

Perhaps most significantly, the 1949 Conventions included Common Article 3—a provision that appeared identically in all four conventions and applied, for the first time, to non-international armed conflicts. While Common Article 3 addressed primarily the treatment of the living—prohibiting violence, torture, and degrading treatment—its presence signaled a recognition that humanitarian principles could not be limited to wars between states.[62]

The Additional Protocols of 1977

The decolonization struggles of the 1950s and 1960s, the Vietnam War, and the proliferation of guerrilla conflicts around the world revealed new gaps in the 1949 framework. Wars of national liberation blurred the line between combatants and civilians. Aerial bombardment and new weapons technologies created casualties far from any recognizable front line. The distinction between international and non-international armed conflicts—so central to the 1949 architecture—seemed increasingly artificial.

In 1977, a diplomatic conference adopted two Additional Protocols to the Geneva Conventions. Protocol I addressed international armed conflicts, while Protocol II extended protections to victims of non-international conflicts meeting certain thresholds of intensity and organization.[63]

For the dead, Protocol I was particularly significant. Article 33 addressed the search for the missing and dead, requiring each party to a conflict to search for persons reported missing by an adverse party. It established the right of families to know the fate of their relatives—a principle that would later be recognized as a fundamental element of international humanitarian law.[64]

Article 34 addressed the remains of deceased persons and the location of graves. It required parties to facilitate the return of remains to the home country upon request and to facilitate access to gravesites by relatives. The article also established protections for gravesites, prohibiting their disturbance except for exhumation conducted in accordance with proper procedures and with respect for the remains.[65]

These provisions reflected a growing recognition that obligations to the dead extended beyond the immediate aftermath of conflict. The dead retained a claim on the living, and so did their families. The right to know, the right to mourn, the right to bring remains home—these were not merely sentimental preferences but legal entitlements recognized by international law.

A Living Framework

The Geneva Conventions and their Additional Protocols remain the foundation of international humanitarian law today. As of 2024, all 196 states worldwide have ratified the four 1949 Conventions, making them among the most universally accepted legal instruments in human history.[66]

Yet the conventions are not static documents. Their interpretation evolves through state practice, the decisions of international tribunals, and the guidance of the ICRC, which serves as the "guardian" of international humanitarian law. The ICRC's commentaries on the conventions—updated in recent years to reflect contemporary developments—provide authoritative interpretation of treaty provisions.[67]

The framework continues to face challenges. Non-state armed groups, which now participate in the majority of armed conflicts worldwide, often lack the capacity or willingness to comply with humanitarian obligations. The rise of terrorism and counterterrorism operations has strained traditional categories of combatants and civilians. And the dead of contemporary conflicts (in Syria, Yemen, Ukraine, and elsewhere) are not always afforded the protections that the conventions promise.[68]

But the aspiration endures. Each revision of the Geneva framework has been written in response to the failures of the previous generation. Each has attempted to close gaps that war had exposed. And each has reaffirmed the fundamental principle that even in the midst of an armed conflict, the dead retain their dignity, and the living retain obligations toward them.

The conventions do not eliminate suffering. They do not prevent atrocity. But they establish a standard against which conduct can be judged and to which conscience can appeal.

They are, in the end, a promise.

That we will try to remain human, even when humanity seems lost.

That we will honor the dead, even when we cannot save the living.

That we will remember, even when forgetting would be easier.

Artifact: The POW Information Card (1914–Present)

"A rectangle of paper. A life reduced to categories. A family's only thread to the missing."

During the First World War, the International Prisoners of War Agency in Geneva processed approximately five million index cards, each one representing a captured, wounded, or dead soldier. The cards were standardized: name, rank, nationality, date and place of capture, state of health, next of kin. Staff transcribed information from handwritten lists onto these cards, filed them alphabetically, and cross-referenced them against incoming inquiries.

The cards themselves were modest—a few inches of paper, typed or handwritten, stamped with the Red Cross emblem. But for families who had received no word in months, a single card could mean everything. It could confirm a son was alive. It could reveal where a husband was held. It could, in the worst cases, provide the date and location of a death—closure, however painful.

These cards still exist. Millions are archived in Geneva, now digitized but still bearing the marks of the hands that filled them. Some are smudged with mud or marked with corrections. Some contain errors—names misspelled, dates uncertain. But each one represents an attempt to keep a person from disappearing into the machinery of war.

The POW card is not a monument. It is a bureaucratic form. And yet, in its quiet way, it embodies the promise of humanitarian law: that even in captivity, even in death, a person has a right to be known. A family has a right to be told. A name has a right to survive.

Today, the ICRC's tracing systems have evolved—digital databases, DNA matching, satellite imagery. But the principle remains the same. Somewhere, a card is still being filled out. A name is still being recorded. A family is still waiting for the thread that connects them to someone lost.

Part C: Law in Practice and Its Limits-Implementation, Institutions, and the Forgotten

"A law unlearned is a law unenforced."

From Law to Practice: Teaching the Rules of War

The Geneva Conventions are among the most widely ratified treaties in human history. Yet ratification is not implementation. A soldier who has never heard of Article 17 cannot honor it. A commander who does not understand the distinction between combatant and civilian cannot protect either. Law, to be effective, must be taught.

This is the work of International Humanitarian Law (IHL), the body of rules that seeks to limit the effects of armed conflict. IHL does not ask whether war is just. It asks only: given that war exists, how do we preserve what remains of our humanity? The answer, it turns out, requires education.

Military academies now include IHL in their curricula. The International Committee of the Red Cross trains armed forces around the world in the laws governing the treatment of prisoners, the wounded, and the dead. National Red Cross and Red Crescent Societies, including the American Red Cross, conduct outreach programs that bring these principles to soldiers, students, and civilians alike. The teaching is not abstract; it is grounded in scenarios, in case studies, in the recognition that every rule exists because someone, somewhere, suffered its absence. To teach IHL is to acknowledge that laws were written in blood. Those laws must be learned so that less blood is spilled.

And yet, the teaching is fragile. Curricula can be cut, and training can be rushed. In the chaos of deployment, the rules may be forgotten or overridden by commanders who see them as obstacles. The gap between law and practice is never fully closed. This is why the work continues. Why volunteers and educators and advocates return again and again to the same principles, the same conventions, the same stubborn insistence that even in war, there are limits. The law exists. But it lives only when it is learned.

I know this work from the inside. As a volunteer with the American Red Cross, I have taught International Humanitarian Law. I have walked through the legal frameworks designed to protect the dignity of the dead. I have helped families navigate the particular grief of wartime loss. And through Restoring Family Links, I have witnessed the search for the missing—the hope, the waiting, the silence that sometimes answers. This chapter is not history observed from a distance. It is history I have been asked to carry forward.

The ICRC Today: From Tracing to Identification

"The work did not end when the treaties were signed. It began."

International humanitarian law provides the framework. But frameworks do not search for the missing. They do not match bones to names. They do not deliver the news that a family has waited decades to hear.

That work falls to human hands. For more than a century and a half, those hands have belonged largely to the International Committee of the Red Cross and its global network of National Societies. What began as an improvised effort to connect wounded soldiers with their families has evolved into the world's largest humanitarian network dedicated to restoring family links and accounting for the missing and dead.

Origins of the Central Tracing Agency

The ICRC's tracing work began not by design; it came from desperate necessity.

During the Franco-Prussian War of 1870–1871, the ICRC established an informal information bureau in Basel, Switzerland, to transmit news about prisoners and wounded soldiers between French and German families.[69] The effort was modest—a handful of volunteers processing letters and lists—but it established a precedent that would prove transformative.

By the First World War, this improvised system had grown into the International Prisoners of War Agency, headquartered in Geneva. Operating from the Musée Rath, a converted art museum, the agency processed an astonishing volume of information. Over the course of the war, it handled approximately five million index cards, transmitted millions of messages, and responded to countless inquiries from families desperate for news of their loved ones.[70]

The agency's methods were painstakingly manual. Information arrived from belligerent powers in the form of handwritten lists, typed rosters, and official notifications. Staff members—many of them volunteers—transcribed this information onto standardized cards, filed alphabetically by name. When an inquiry arrived, staff would search the files by hand, cross-referencing names, units, and dates in hopes of finding a match.[71]

The work was tedious, often heartbreaking, and never sufficient. Families wrote pleading letters. Photographs arrived of young men in uniform and children asking after fathers. Many inquiries could not be answered. The dead did not always leave records, and the missing did not always become found.

Yet the agency persisted, and its work saved countless families from the torture of unknowing. A telegram confirming a prisoner's location. A letter forwarded across enemy lines. A death certificate, however painful, that allowed mourning to begin. These were not mere administrative services. They were lifelines.

After the war, the agency was renamed the Central Prisoners of

War Agency, and its role was formalized in the 1929 Geneva Convention. When the Second World War erupted, the agency—by then known simply as the Central Agency—scaled up to meet an even greater challenge.[72]

The Central Tracing Agency in World War II and Beyond

The Central Agency's work during the Second World War dwarfed all previous efforts. By the war's end, the agency had compiled approximately 45 million index cards and processed some 120 million messages between prisoners, internees, and their families.[73] The card index alone occupied an entire floor of a Geneva building, staffed by hundreds of employees working in shifts.

The agency also confronted horrors that no previous conflict had produced. As Allied forces liberated concentration camps, the ICRC received fragmentary lists of survivors—and far longer lists of the dead. Families who had received no news for years now learned, often through agency channels, that their relatives had perished in the Holocaust. The task of tracing became inseparable from the task of documenting atrocity.[74]

After 1945, the agency's work did not diminish—it transformed. The postwar period brought new categories of missing persons: refugees separated from families during flight, displaced persons scattered across occupied zones, children orphaned by bombing or deportation. The Cold War added further complications, as political divisions prevented the free flow of information across the Iron Curtain.[75] In 1960, the Central Agency was renamed the Central Tracing Agency (CTA), reflecting its expanded mandate beyond prisoners of war to include all persons missing as a result of armed conflict, internal violence, or natural disaster.[76] The name change signaled the broader evolution of the ICRC from merely an intermediary between belligerent states to a global humanitarian organization addressing the full spectrum of family separation.

Restoring Family Links: A Global Network

Today, the ICRC's tracing and family reconnection work operates under the banner of "Restoring Family Links" (RFL), a global network that encompasses the ICRC, 191 National Red Cross and Red Crescent Societies, and their International Federation.[77]

The RFL network handles an extraordinary range of cases. Families separated by armed conflict in Syria, Yemen, and South Sudan. Migrants who have lost contact with relatives during dangerous journeys across the Mediterranean or through Central America. Survivors of natural disasters seeking news of loved ones. Elderly individuals searching for siblings separated during the partition of India or the Korean War. The mandate is broad: to restore contact between family members separated by conflict, disaster, or migration, and to clarify the fate of those who are missing.[78]

The scale of the work is immense. In 2022 alone, the RFL network collected over 47,000 tracing requests, facilitated the exchange of more than 260,000 Red Cross messages, and helped reunite or restore contact between thousands of separated family members.[79] Behind each statistic lies a human story: a mother reconnected with a son she had not seen in twenty years; a refugee who finally learns that his brother survived; a family that receives confirmation, however painful, that their search has ended.

The RFL network employs multiple methods to fulfill its mandate:

Red Cross Messages: In contexts where normal communication is impossible—conflict zones, detention facilities, areas without telecommunications infrastructure—the Red Cross Message service allows families to exchange brief, personal messages through ICRC and National Society channels.

Messages are screened to ensure they contain only family news, then hand-delivered by delegates or volunteers.[80]

Tracing Requests: Relatives can submit a formal tracing request through their National Society to find missing family members. The request triggers a search through ICRC databases, inquiries to authorities in relevant countries, and field investigations by ICRC delegates. The process can take months or years, and not all cases are resolved—but many are.[81]

Family Reunification: In cases where separated family members are located alive, the RFL network facilitates reunification. This may involve arranging travel documents, coordinating with immigration authorities, providing travel assistance, or simply enabling a phone call or video connection across continents.

Certificates and Attestations: For families who need official documentation (death certificates, attestations of detention, confirmation of family relationships) the ICRC can provide certificates based on its records. These documents often prove essential for legal proceedings, inheritance claims, or simply for the peace of mind that comes with official acknowledgment.[82]

The Forensic Turn: Science in Service of the Missing

The late 20th century brought a revolution in the search for the missing: forensic science.

The development of DNA identification techniques in the 1980s and 1990s greatly improved the likelihood of identifying human remains. Where traditional identification methods (personal effects, dental records, physical characteristics) often failed, especially when remains were fragmentary or decomposed, DNA analysis offered the prospect of definitive answers. A bone fragment, a tooth, a lock of hair—any biological sample containing intact DNA could potentially be matched to a reference sample from a family member.[83]

The ICRC was not slow to recognize the implications. Beginning in the 1990s, the organization developed expertise in forensic identification and began partnering with national authorities, academic institutions, and specialized organizations to support the recovery and identification of the missing.[84]

The Balkans provided a tragic proving ground. The wars that accompanied the dissolution of Yugoslavia produced tens of thousands of missing persons—victims of massacres, ethnic cleansing, and displacement. Mass graves dotted the landscape of Bosnia, Croatia, and Kosovo. Families demanded answers. And slowly, painstakingly, the forensic community began to provide them. The International Commission on Missing Persons (ICMP), established in 1996 at the initiative of U.S. President Bill Clinton, pioneered the large-scale application of DNA identification to conflict-related missing persons.[85]

Working in partnership with the ICRC and national authorities, the ICMP developed protocols for the collection of reference samples from families, the recovery and analysis of remains from mass graves, and the management of vast DNA databases. By 2023, the ICMP had assisted in the identification of more than 20,000 individuals from the Balkan conflicts alone.[86]

The ICRC's role in this forensic ecosystem is primarily one of support and coordination rather than direct laboratory work. The organization provides guidance on the proper handling of human remains in conflict zones, trains national authorities and humanitarian workers in best practices, and advocates for the legal and institutional frameworks necessary for effective identification programs.[87]

In 2003, the ICRC published *The Missing: ICRC Action to Resolve the Problem of People Unaccounted for as a Result of Armed Conflict or Internal Violence*, a comprehensive framework addressing the legal, humanitarian, and forensic dimensions of the missing persons problem.[88] The document emphasized that the right of families to know the fate of their relatives was not merely a humanitarian concern but a legal obligation under international humanitarian law.

The ICRC has also developed specific guidance on forensic best practices. Its 2009 publication *Missing People, DNA Analysis and Identification of Human Remains* provides detailed protocols for the collection, storage, and analysis of biological samples, as well as guidance on the ethical and legal issues that arise in forensic identification. Updated guidance continues to address emerging technologies and evolving best practices.[89]

Coordination with National Authorities

The ICRC does not work in isolation. Effective accounting for the missing requires coordination with national authorities, military organizations, and specialized agencies around the world.

In the United States, the primary agency responsible for accounting for missing service members is the Defense POW/MIA Accounting Agency (DPAA), established in 2015 through the merger of several predecessor organizations.[90] The DPAA conducts research, investigation, and recovery operations worldwide, seeking to account for the more than 81,000 Americans still missing from past conflicts—the majority from World War II and the Korean War.[91]

The ICRC maintains regular dialogue with the DPAA and similar national agencies, sharing information and coordinating efforts where appropriate. When ICRC delegates operating in conflict zones encounter information relevant to missing Americans—or missing persons of any nationality—protocols exist for transmitting that information to the appropriate authorities.[92]

National Red Cross and Red Crescent Societies also play crucial roles. The American Red Cross, for example, operates a Holocaust and War Victims Tracing Center that assists families searching for information about relatives affected by World War II and subsequent conflicts.[93] Similar programs exist within National Societies worldwide, each adapted to the specific historical and contemporary needs of their populations.

International coordination also extends to multilateral frameworks. The ICRC participates in the International Conference of the Red Cross and Red Crescent, which brings together states party to the Geneva Conventions and components of the Red Cross and Red Crescent Movement to address pressing humanitarian concerns. Resolutions adopted by the International Conference have repeatedly emphasized the importance of accounting for the missing and supporting the families of those who remain unaccounted for.[94]

The Human Dimension

Behind the legal frameworks, the forensic protocols, and the institutional structures lies the simple human reality that families who do not know what happened to their loved ones live in a particular kind of anguish.

The psychological literature documents this experience under various names: ambiguous loss, frozen grief, the impossibility of mourning without a body to bury or a grave to visit.[95] Families of the missing often describe feeling suspended between hope and despair, unable to move forward, unable to let go. The absence of information is not merely frustrating—it is existentially disorienting.

The ICRC's work addresses this human dimension directly. Tracing is not merely an administrative function but a form of care. The Red Cross message that reaches a detention facility is not merely a piece of paper but proof that someone outside still remembers, still searches, still loves. The identification of remains after decades is not merely a forensic achievement but a restoration of dignity to the dead and peace to the living.

ICRC delegates who work on missing persons cases often speak of the profound privilege and burden of their work. They are the bearers of news that families have awaited for years—sometimes news of reunion, sometimes news of death. They witness grief in its rawest forms. They also witness resilience, gratitude, and the remarkable human capacity to find closure even in tragedy.[96] The organization has increasingly recognized the need to support not only the informational but the psychosocial needs of families of the missing. Programs now exist to provide psychological support, facilitate peer connections among affected families, and assist with the practical legal, economic, and social challenges that accompany the disappearance of a family member.[97]

THE WORK CONTINUES

More than 160 years after Henri Dunant organized villagers to care for the wounded at Solferino, the ICRC continues its mission. The specific challenges have evolved—from battlefield casualties to concentration camp victims to migrants drowned in the Mediterranean—but the fundamental commitment to restore family links, to account for the missing, and to ensure that the dead are not forgotten remains.

The work is never finished. Every armed conflict produces new missing persons. Every humanitarian crisis separates families. Every year, the ICRC opens thousands of new tracing cases and closes others—some with joyful reunions, some with final confirmations, some with the notation "fate unknown" that represents the limits of human knowledge.

The work continues because it must. Because the right to know is not merely a legal abstraction but a human necessity. Because families cannot fully mourn those whose fates remain uncertain. Because the dead deserve to be named, and the living deserve to know where they lie.

The ICRC's tracing work is, in the end, an extension of the same impulse that drove Dunant to organize care for the wounded at Solferino: the recognition that human suffering demands a human response. Law provides the framework. Forensic science provides the tools. But the work itself is an act of solidarity, a refusal to allow the chaos of conflict to sever the bonds that connect us to one another.

To search for the missing is to insist that they still matter.

To restore family links is to affirm that connection survives even war.

To identify the dead is to give them back their names.

And in giving them back their names, we reclaim something of our own humanity.

The Cost of Disregard: When the Dead Are Denied

"When the dead are denied burial, a deeper violence occurs. The violence of forgetting."

There is a wound deeper than death.

It is the wound of being discarded.

When we speak of burial law as rite and as a moral act, we must also speak of what happens when it is withheld. The denial of burial is not simply neglect; it is an erasure of identity, a rupture of dignity, a rupture of memory. Across time, this denial has been used as punishment, propaganda, and policy. And the consequences reverberate long after the bodies are gone.

In genocide, mass graves are not merely byproducts of violence, they are part of the design. They conceal names. They sever family ties. They create what scholar Alexander Hinton calls "zones of social death," places where both physical bodies and cultural legacies are buried beneath silence.[98]

During the Holocaust, millions were incinerated or dumped into pits without names, without markers, without mourning. The Nazis deliberately mechanized death to eliminate not only life but memory. Sites like Babi Yar, a ravine outside Kyiv where over 33,000 Jews were murdered in two days, remain stark reminders of what happens when the dead are stripped of identity.[99] Even now, many Holocaust victims remain unnamed, their lives reduced to ash and absence.

In Cambodia, under the Khmer Rouge regime (1975–1979), over a million people were executed and buried on what became known as the "killing fields." Shallow graves, mass pits, and bone-strewn clearings dot the countryside.[100] The regime forbade mourning rituals. Families were not allowed to grieve. In the eyes of the state, the dead were not to be remembered—they were to be erased.

The Rwanda Genocide (1994) followed a similar pattern. Bodies were dumped in rivers, piled in churches, buried hastily or not at all. Entire communities were destroyed, and with them, the cultural rituals that might have made mourning possible.[101] Survivors speak not only of grief, but of the trauma of being unable to bury their dead, of fleeing with their memory unanchored, their sorrow unfinished.

Modern warfare, too, has often forsaken the dead. In the era of drone strikes and remote combat, many bodies are never recovered. Civilians killed in missile attacks or insurgents left behind after raids frequently go unburied or remain unidentified. In places like Syria and Yemen, the dead lie in rubble while the world debates legalities.[102]

To leave a body unburied is not just a humanitarian failure. It is a political act. It communicates that the life once held in that body was unworthy of closure, of ritual, of remembrance. It weaponizes silence.

This is why mass graves, when uncovered, provoke such visceral horror. From the killing fields of Srebrenica to the trenches of Katyn, the unearthed bones speak louder than words. They tell us what has been hidden. They testify against forgetting. And they beg for closure, not just for the dead, but for the living.[103]

Forensic anthropologists now play a central role in post-conflict justice. In places like Argentina, Guatemala, and Bosnia, forensic teams have painstakingly exhumed mass graves, cataloging remains, identifying the missing, and returning bones to families decades after death.[104] Beyond scientific work, this is sacred labor, a reconstruction of dignity one bone at a time.

The spiritual cost of denying burial cannot be overstated. In many cultures, a soul that is unburied is a soul that cannot rest. In Yoruba cosmology, the dead who are not properly mourned become restless spirits. In traditional Chinese belief, an unburied ancestor may cause misfortune or disease.[105] Even in secular societies, the image of a body abandoned—on a battlefield, in a ditch, beneath rubble—cuts to the core of what we believe about humanity.

And yet, some of the most grievous acts of disregard happen not in war, but in peace.

In the United States, countless bodies of enslaved people lie in unmarked graves, on plantations, beneath city streets, beneath universities. In New York City, construction workers unearthed a colonial-era burial ground containing over 15,000 remains of free and enslaved Africans, buried without ceremony or stone.[106] These are not ancient histories. They are recent omissions, erased not by violence, but by indifference.

The same is true for many Indigenous communities, whose burial grounds have been desecrated, displaced, or paved over in the name of progress. The children who died in residential schools across North America were often buried in unmarked plots, their families never notified.

Only now, in the 21st century, are these truths being exhumed.[107]

What links these disparate acts, genocide, slavery, colonialism, modern war, is the belief that some lives are less worthy of grieving than others. Philosopher Judith Butler writes that to grieve is to recognize a life as having value. To withhold burial is to deny that value.[108]

And so, the cost of disregarding the dead is never just about bodies.

It is about memory denied. Justice postponed. Humanity lost.

But it is not irreversible.

Every time we unearth a name, rebury a forgotten body, erect a marker where there was once none, we begin to undo the violence. We begin to say: you were here. You mattered. You are not forgotten.

Because burial is not just the closing of life. It is the beginning of remembrance.

And remembrance, when done rightly, restores us all.

The Limits of Law: Forgotten and Marginalized Dead

"Law remembers the victors. It often forgets the rest."

For all their strength, international laws and military codes are not immune to failure. They are written by nations, interpreted by governments, and enforced—when convenient—by power. No matter how comprehensive the Geneva Conventions may be, they have never fully protected the most vulnerable victims of war. Laws codify the ideal. History tells us who gets left out.

The graveyards of modern warfare reveal a troubling pattern: those most likely to be buried without name, ceremony, or legal recognition are often the poor, the colonized, the enslaved, the civilian, the woman, and the enemy who cannot speak for themselves.

In the aftermath of World War I, the British and Commonwealth governments created military cemeteries with white, uniform headstones—symbols of equality in death. But equality was often illusory. In 2021, a report commissioned by the Commonwealth War Graves Commission revealed that over 100,000 African, Indian, and Egyptian soldiers who died in the war were either not commemorated at all or were buried in mass graves without markers.[109] Their names were not lost. They were ignored.

In the U.S. Civil War, newly formed national cemeteries honored Union soldiers with carefully arranged plots and white marble headstones. But Black soldiers were frequently buried in segregated sections, if acknowledged at all.

Confederate cemeteries, meanwhile, erected monuments to generals and rank-and-file soldiers alike—memorials steeped in mythology, silence, and racial exclusion.[110]

During World War II, Japanese-American soldiers in the 442nd Infantry Regiment—the most decorated unit in U.S. military history—fought and died for a country that was simultaneously incarcerating their families in internment camps.[111] Many of their dead were repatriated and buried, but rarely with the same fanfare or narrative prominence as their white counterparts.

The enemy dead are another category often pushed to the periphery of memory. In the wake of battles across the Pacific and European theaters, Axis soldiers were hastily buried in mass trenches or left to decay where they fell.[112] Though international law permitted honorable burial, the emotional fury of war and postwar politics frequently overrode such provisions.

But the marginalization of the dead is not limited to soldiers.

Civilians, especially women and children, make up the majority of casualties in modern war, yet they are the least likely to be buried with dignity. In Rwanda, Syria, Iraq, and Ukraine, entire communities have been bombed into silence. Their bodies lie in collapsed buildings, roadside pits, and unmarked fields.[113] Their deaths are often uncounted, their remains undocumented.

Women in war, especially those who served in noncombat roles or as resistance fighters, have historically faced burial without military honors, without ceremony, and often without record. Female medics, couriers, spies, and partisans were essential to the machinery of war but were frequently excluded from postwar remembrance.[114]

Even in the modern era, disparities persist. In Arlington National Cemetery, concerns have been raised about eligibility, space, and who receives full honors. Civilian contractors who died supporting military operations, Indigenous scouts who served the U.S. Army in the 19th century, and LGBTQ+ veterans dishonorably discharged under now-defunct policies all face uneven legacies in death.[115]

There are also the "invisible dead," those buried under urban development, military waste sites, or forgotten in foreign fields. In 2009, construction on a school in Libya unearthed a mass grave of Italian soldiers from WWII, abandoned for over six decades.[116] No repatriation, no names, no history until the bones forced remembrance.

These limits of law are not merely bureaucratic errors. They reflect the values and blind spots of the societies that write and enforce them.

The question becomes: who is worth remembering?

Law can establish the right to burial, but only memory can enforce it. This is why history must intervene to uncover what law has omitted. To record what politics has erased. To elevate those voices and bodies left in the margins.

And so this book continues—not only to recount the structures we built to honor the war dead, but to shine light on those denied those structures.

Because the true measure of remembrance is not how we bury the famous.It is how we remember the forgotten.

Law as a Voice for the Voiceless

"Where the dead cannot speak, law must speak for them. And where law fails, we must remember louder."

Burial, once a gesture of kin and kind, became law not to regulate grief, but to protect dignity. It was not born from bureaucracy, but from loss. From the sight of bodies left in fields. From the weight of silence after slaughter. From the cries of mothers, chaplains, and gravediggers who knew that even the nameless deserved a name.

The codification of death—the Lieber Code, the Geneva Conventions, the Hague Accords—was an attempt to cling to something human in the inhumanity of war. A voice for the voiceless. A framework to say that this, too, matters. Even now. Even here.

But as we've seen, law does not always reach the margins. It does not always climb the ranks of powerlessness. It does not always bend for the civilian, the colonized, the enemy, the erased. That is the paradox of burial law: it speaks most clearly when applied equally, and yet is most often distorted by the very forces it seeks to tame—politics, prejudice, power.

Still, the aspiration matters. It matters that we wrote these laws. It matters that we debated them, ratified them, translated them into battlefield handbooks and military training manuals. It matters that we tried. Because laws are not only about enforcement. They are also about memory. They are our attempt to set grief to paper, to make mourning legible, to name what must not be forgotten.

And so, even when imperfect, the laws governing burial become a form of resistance. Against anonymity. Against silence. Against oblivion. Against the idea that some bodies are more disposable than others.

They offer a voice—not always loud, not always just, but a voice nonetheless—for those who can no longer speak.

This chapter began with chivalric rituals and sacred rites, traced their evolution into codes of conduct and international law, and ended with the recognition that laws have limits—but remembrance does not.

We are their voices now.

Those who bury, those who write, and those who remember become the caretakers not only of the dead, but of the living legacy they leave behind.

The law may codify death, but it cannot carry the weight of it alone. That burden falls to hands and hearts—to those who dig, who document, who dare to look upon what war leaves behind. Burial statutes may define what ought to happen after death, but only people, flawed, faithful, and often forgotten, determine what actually does.

We will meet them soon. In the wheat fields of Antietam and the muddy roads of Shiloh, we will find stretcher-bearers, chaplain-gravediggers, contraband laborers, and nurses kneeling beside the dead with pencil and paper. Their names—some remembered, many not—will guide us into the next chapter.

The Civil War did not just fracture a nation. It awakened it to the logistics of loss.

Interlude: Between the Laws and the Losses

"The rules were written. The graves were marked. And yet, the war came anyway—and it came with more dead than we had words for."

There is a space between what is promised and what is practiced.

The sacred codes of burial, etched in clay, scribed in scripture, and finally typed into treaties, offer dignity in theory. But the battlefield offers only blood, mud, and delay.

Before the Civil War, Americans buried their dead in churchyards, family plots, and village greens. Death was local, tended to by those who knew your name. Grief was shared under one roof, beside one table.

But the war changed that.

It brought death that traveled. Death by rail, death by mass grave, death that never came home. It brought letters that read "body not recovered." It brought lists. Numbers. And questions that the law alone could not answer.

What happens when the body is too torn to identify? When a uniform is the only evidence of a life? When bones are gathered into barrels, and names vanish into columns?

This is where we now walk.

Between the parchment and the pit. Between the code and the corpse. Between ancient ritual and industrial ruin.

The next chapter begins not with legislation, but with shovels.

Artifact: Antigone's Defiance (c. 441 BCE)

A Broken Burial—The Law of Love vs. the Law of the State

Sophocles' *Antigone* is one of the earliest and most enduring testaments to the sacred obligation of burial. When King Creon forbids the burial of her brother Polynices, Antigone faces death to honor what she calls the "unwritten laws" of the gods. Her resistance is not political—it is ritual, spiritual, and deeply personal.

The artifact is not a relic but a story, the timeless literary documentation of conscience. It reminds us that long before modern legal frameworks, burial was defended with defiance, protected with poetry, and carried by those who dared to say no to erasure.

Antigone's voice still echoes in every chaplain who blesses an enemy grave. In every family who demands their child not be forgotten. In every gravedigger who asks not whom they bury—but how.

***

Between the parchment and the pit. Between the code and the corpse. Between ancient ritual and industrial ruin. The next chapter begins not with legislation, but with shovels.

Bridge

From Law to Labor

The Weight of Implementation

"The laws were written. The graves were marked. And yet, the war came anyway—and it came with more dead than we had words for."

Part I of this book has traced the foundations of burial as a human act from the pollen-strewn graves of Neanderthals to the diplomatic conferences of Geneva and The Hague. We have seen how the instinct to honor the dead became custom, how custom became code, and how code became international law. We have watched the slow emergence of a principle: that even in war, the dead retain their dignity, and the living retain obligations toward them.

But principles do not bury bodies.

Laws do not dig graves.

Treaties do not wash the blood from a shattered face or write the letter that tells a mother her son will not come home.

That work falls to hands—human hands, trembling hands, hands that have never held a shovel before and will never forget the weight of what they carried.

Part II turns from the architecture of remembrance to the labor of it. We move from the halls of diplomacy to the fields of Antietam, the trenches of the Argonne, the beaches of Normandy, and the jungles of Vietnam. We follow not the lawmakers but the law's servants—the stretcher-bearers and gravediggers, the clerks and chaplains, the freedmen and volunteers, the women who stitched names into cloth, and the men who carved them into wood.

These were the caretakers of America's dead.

They were rarely trained for the work. Most had never seen a corpse before being asked to collect hundreds. They were young farm boys and factory workers, immigrants and freedmen, nurses and nuns. They came from every corner of the country, and they were united only by the task that history assigned them: to ensure that the fallen did not vanish into anonymity.

The American experience of war burial is unique in some ways—shaped by the vastness of the continent, the legacy of slavery, the ideology of democratic equality, and the peculiar circumstances of a nation that has fought most of its wars far from home. But it is also universal. Every army that has ever marched has left bodies behind. Every nation that has ever mourned has faced the question: *who will bury the dead?*

In America, the answer emerged slowly, painfully, through trial and catastrophe. The Civil War, which tore the nation apart, also created its first national cemeteries and its first systematic approach to military burial. The World Wars, which sent millions overseas, demanded a bureaucracy of death that could operate across oceans and continents. Korea and Vietnam, fought in the shadow of nuclear anxiety and political division, tested whether a country could honor its dead when it could not agree on why they had died.

Through it all, the caretakers remained. They were not always honored, and they were not always seen. Their names rarely appear in the histories of battles and campaigns. But without them, there would be no rows of white crosses, no walls of names, and no tombs of unknowns. Without them, the dead would have been swallowed by the earth without witness.

This part of the book is their story.

It begins in 1861, when a nation that had never faced mass death on its own soil suddenly found itself drowning in it. The systems that arose from that grief—imperfect, improvised, often unjust—would shape how Americans understood their obligations to the fallen for generations to come.

The laws had been written.

Now came the labor.

PART II

AMERICA'S DEAD AND THEIR KEEPERS

"The country that sends its young to die must decide how it will remember them."

The Civil War forced America to confront something it had never seen at such scale: mass death on its own soil. The systems that arose from that grief—cemeteries, ledgers, burial teams—would become the framework for how the United States handled its war dead for generations to come.

This part of the book traces the evolution of those systems and the human cost behind them. From the battlefield burial crews of the Civil War to the trench diggers of World War I, from the segregated units of World War II to the haunted jungles of Vietnam, it follows the people who took on the sacred duty of remembering the fallen.

These were not generals or diplomats. They were clerks. Gravediggers. Mothers. Farmers. Often poor, often young, and often Black or marginalized, they became the caretakers of a nation's grief.

They worked in silence. But their labor echoes still.

Reflection

Clara Barton founded the American Red Cross at age fifty-two. I was fifty-two when I read that fact and felt something shift inside me. I had spent years believing that my significant contributions were behind me—that whatever I was going to accomplish, I had already accomplished. The decades ahead seemed like a denouement, not a beginning. But Clara Barton was just getting started at fifty-two. She had already nursed soldiers on Civil War battlefields, already established the Office of Missing Soldiers, and already traveled to Europe and witnessed the birth of the International Red Cross. And then, at an age when many people slow down, she founded an organization that would serve millions for the next century and a half. She did this in a long dress with petticoats and a corset. She did this in a world that did not believe women belonged in public life. She did this despite opposition, illness, and setbacks that would have stopped most people.

If she could do it, I thought, I could write this book.

My work with the Red Cross has shown me how important Clara Barton and Henri Dunant were—not just in founding organizations, but in establishing the principle that care must accompany conflict.

The wounded must be tended.

The dead must be honored.

The missing must be sought.

The Civil War produced horrors that forced a reckoning—and from that reckoning emerged systems of care that persist today. I am proud to continue that legacy through my volunteer work.

And I am grateful to Clara Barton for showing me that it is never too late to begin.

CHAPTER THREE

THE CIVIL WAR: MOURNING A NATION TORN

PART A: THE CRISIS

"The North had the factories. The South had the fields.
But it was grief that built the cemeteries."

Interlude: The Shovel and the Nation

They did not plan for death like this—not in rows of thousands, not in trenches dug with broken picks, not in the arms of strangers far from home.

Before the war, Americans buried kin by hand and by hymn. But this war asked something new of the country—To bury the unkno wn.To mourn en masse.To dignify the fallen without name, without preparation, without end.

And so the country improvised its rituals. In that improvisation, a new kind of nation emerged—one bound by memory, not just victory.

A Nation Unprepared: The Dead Pile Up

"The war made cemeteries of cornfields."—Civil War chaplain

When the American Civil War broke out in 1861, the nation was utterly unprepared—not just for the scale of combat, but for its consequences. There was no standing infrastructure for war dead. No quartermaster divisions assigned to burial, no federally managed cemeteries, no standard identification system. The country had never confronted mass death on its own soil, and certainly not at the scale the Civil War would bring.

And so the dead piled up.

On battlefields like Antietam, Shiloh, and Gettysburg, soldiers fell by the thousands in the span of hours. In the days that followed, their bodies were left in wheat fields, orchards, and along riverbanks, often exposed for days or weeks. Horses trampled corpses. Rain and summer heat bloated them beyond recognition. Bones surfaced months later during plowing.[1]

Initial burials were crude, desperate, and highly localized. Fellow soldiers or nearby civilians dug shallow graves with bayonets or wooden planks. Crosses were made from fence posts, or names scratched into bark or boards. But in many cases, no marker was left at all. The haste of retreat or the chaos of continued fighting meant that many were buried hastily, or not at all. There was no consistent method of identification. Union soldiers sometimes pinned paper tags to their uniforms and wrote their names on scraps of cloth before battle. But such efforts were personal, not official. Confederates, many of whom lacked proper uniforms or tags, were even less likely to be identified.[2]

As the death toll climbed, so did public unease. Newspaper reports began detailing the horrific conditions of battlefield dead. Letters from soldiers and chaplains described the stench of decomposing bodies and the trauma of leaving friends behind.[3] A moral crisis loomed: what civilized nation would allow its fallen sons to rot unburied? The lack of preparation became a national scandal after the Battle of Gettysburg in 1863. Visitors to the field weeks after the fighting found hastily dug graves overturned by animals, bodies still exposed in the summer sun. The public outcry helped spark the federal government's decision to act. But Gettysburg was already one of hundreds of sites where death had overwhelmed local communities.

Before federal intervention began, much of the early burial labor was done by ordinary people—townsfolk, clergy, and often newly freed Black Americans. In the rural communities where battles swept through, it was frequently the poor—tenant farmers, day laborers, and those with no means to flee—who bore the burden. They were sometimes paid, sometimes conscripted, often simply left with no choice. The bodies were on their land, and someone had to dig. Their names appear in no rosters. Their labor earned no recognition. They buried strangers because the army had moved on, and the dead remained. Their names were rarely recorded. Their methods were improvised. Their grief was compounded by isolation.[4]

The Civil War broke many things. Among the most haunting was its rupture of ritual.

Without systems for burial, the war dead became both victims and ghosts: unclaimed, unseen, unresolved. The battlefield became not just a site of military struggle, but of spiritual limbo. And for a nation that had long idealized the "good death," surrounded by family, with last words spoken and a proper burial granted, the reality of anonymous corpses shattered deeply held beliefs about honor and salvation.[5]

From this rupture, a reckoning began. Graves registration, formal cemeteries, identification efforts, and federal policy would all rise from this chaos. But in these early years of war, what mattered most was the human response, the chaplain who said a prayer over strangers, the farm girl who marked a grave with flowers, the soldiers who dug holes in frozen ground with numb hands.

They did not yet know they were part of history.

They only knew the dead could not be left behind.

The Dead at Sea: Naval Warfare and the Waters That Hold No Graves

The Civil War was not only fought on land. It was fought on rivers and coastlines, in harbors and open water, aboard vessels that carried death as surely as any battlefield. Naval warfare presented a unique challenge for the care of the dead: there was no ground to dig.

When sailors died aboard ship—from combat, disease, or accident—their bodies could not be preserved for long. Without refrigeration, in the heat of Southern waters, decomposition began within hours. Captains faced an impossible choice: keep the dead aboard, risking disease and destroying morale, or commit them to the sea.[6]

The ritual of burial at sea was ancient, but the Civil War industrialized it. Ironclad battles like Hampton Roads produced casualties that had to be addressed quickly. Blockade duty, which kept ships at sea for months, meant that sailors who died could not be returned to port for burial. The ocean became their cemetery.[7]

The ceremony was simple but sacred. The body was sewn into canvas, weighted with shot or chain to ensure it would sink. A chaplain, if present, read from the burial service. The ship's company assembled on deck. And then the body was committed to the deep, sliding from beneath the flag into waters that would hold it forever.[8]

For families at home, this meant there was no grave to visit and no headstone to tend. Only a set of coordinates, if those were even recorded, marking the place where a son or husband had been given to the sea.

The USS *Monitor*—the famous ironclad that battled the *CSS Virginia*—sank in a storm off Cape Hatteras in December 1862, taking sixteen sailors with her.[9] Their bodies remained entombed in the wreck for over a century, until the ship was raised and two unknown sailors were recovered and eventually buried at Arlington National Cemetery in 2013.[10] Even then, more than 150 years after their deaths, the effort to bring them home continued.

The sea claims the dead, but it does not release them easily.

For the caretakers of naval dead—the shipmates who sewed the canvas, the officers who read the prayers, the sailors who watched their friends disappear beneath the waves—burial at sea carried its own particular grief. There was no marker to return to. No ground to kneel beside. Only memory, and the knowledge that somewhere beneath the surface, the dead waited.

The Ghosts of Andersonville: When Identity Became a Crisis

"Thirteen thousand graves. Thirteen thousand questions."

No site embodied the war's crisis of identification more starkly than Andersonville. Camp Sumter, as it was officially known, was a Confederate prison-of-war camp in southwest Georgia. Opened in February 1864, it was designed to hold 10,000 Union prisoners. By

August, it held over 32,000, crammed into a 26-acre stockade without adequate shelter, food, sanitation, or medical care. Men drank from the same stream that served their latrine. They died of dysentery, scurvy, exposure, and starvation at a rate of over 100 per day. By the time the war ended, nearly 13,000 Union soldiers had died at Andersonville, more than at any single battle of war. And almost none of them had been properly identified.

The dead were buried in trenches outside the stockade walls, laid shoulder to shoulder in rows, with only numbered wooden stakes to mark their positions. Confederate recordkeeping collapsed under the weight of death. Names were lost. Units were confused. Thousands of men simply vanished into Georgia clay.

For the families who had received no word—no telegram, no letter, no body—Andersonville became a wound that would not close.

The work of healing that wound fell to an unlikely team: Clara Barton, already famous for her nursing work and her Office of Missing Soldiers, and Dorence Atwater, a former prisoner who had survived Andersonville by serving as a clerk in the camp's hospital. Atwater had secretly copied the death register—a list of names, numbered to correspond with the grave stakes—and smuggled it out when he was released in a prisoner exchange.

In the summer of 1865, Barton and Atwater traveled to Andersonville with a War Department expedition. For weeks, they worked through the heat and the stench of the burial trenches, matching Atwater's list to the numbered stakes, replacing wooden markers with permanent headboards, and compiling the most complete record of the dead that would ever exist.

They identified 12,912 of the 12,920 graves. Only 460 remained unknown.

This remarkable achievement was one of the most successful mass identification efforts of the 19th century. And it was accomplished not by the army, not by the government, but by a woman with a list and a man who had refused to let his comrades disappear.

Andersonville became the first national cemetery dedicated to prisoners of war. Though her partnership with Atwater would later sour over credit and compensation, Barton returned to Washington a national hero, and the significance of their work endured. They had proven that the dead could be named. That even after the systems failed, after the records collapsed, after hope itself seemed buried, someone could still reach into the earth and pull back a name.

It would not bring the dead home.

But it would tell their families where they lay.

And for thousands of mothers, wives, and children, that was enough.

Part B: The Response-The Birth of National Cemeteries

"The dead had given their last full measure. The nation owed them more than silence."

Before the Civil War, the United States had no national cemeteries. Military dead were buried wherever they fell, sometimes repatriated by families, but more often consigned to unmarked graves far from home. The idea that the government should honor its fallen with official burial grounds did not yet exist.

That changed in the face of overwhelming death.

In 1862, amid mounting casualties and growing public distress, President Abraham Lincoln signed legislation authorizing the creation of national cemeteries.

The first designated cemetery under this new policy was established in Sharpsburg, Maryland, near the site of the Battle of Antietam, the bloodiest single-day battle in American history.[11] Others quickly followed at Stones River, Chattanooga, Gettysburg, and dozens of other key sites.

These cemeteries were not just practical. They were political and spiritual declarations that signaled a shift in the relationship between the federal government and its citizens, particularly its soldiers. For the first time, the state assumed responsibility not just for deploying military force, but for caring for those who had died in its service. As historian Drew Gilpin Faust writes, this was the birth of "a republic of suffering" in which death became a shared national burden.[12]

The design of these early cemeteries reflected republican ideals. Soldiers were buried in orderly rows regardless of rank, an egalitarian gesture in death, even if not always observed in life. Each grave was to be marked, cataloged, and remembered.[13]

But the process was neither immediate nor inclusive. Initially, national cemeteries were established only for Union dead. Confederate soldiers, viewed as rebels rather than fallen Americans, were excluded from federal burial grounds. Many were buried in mass graves or by their own communities, without the funding or oversight afforded to Union dead.[14] This policy would become one of the most enduring wounds of the war's aftermath, shaping how memory was politicized in the decades to come.

Additionally, Black soldiers, though technically buried in national cemeteries, were sometimes interred in segregated sections. Even in death, the racial hierarches of the living endured.[15] It was not until much later that formal efforts were made to rectify these disparities in how Black service members were memorialized.

The work of building these cemeteries fell to soldiers, local laborers, and newly freed Black Americans, many of whom had previously worked as gravediggers or field hands. Their names rarely appear in official records, but their labor shaped the landscapes of remembrance we now consider sacred.[16]

By the end of the war, fourteen national cemeteries had been established. That number would grow to over seventy in the decades that followed. These spaces became hallowed grounds, not only for the dead, but for the living in search of closure.

They also became the sites of national identity-making.

In 1864, the federal government took control of the grounds of Arlington House, once the estate of Confederate General Robert E. Lee, and began burying Union dead there—both a logistical decision and a symbolic one.[17] Arlington National Cemetery would become the most iconic of all U.S. military cemeteries, a place where the politics of death and reconciliation unfolded visibly.

The birth of the national cemetery system marked a transformation. No longer would soldiers be buried as isolated individuals, forgotten in the wilderness of war. Now, they would lie among comrades, under the care of a grateful republic.

The dead had made the ultimate sacrifice.

The living, at last, had begun to answer it.

The Power of Paper: Ledgers, Names, and Records (Clerks, Clergy, and the Burden of Naming)

"The war wrote story in blood. We tried to rewrite it in ink."—Army clerk, 1866.

Even in war, someone reaches for the pen.

Behind the sounds of musket fire and field hospitals, another kind of caretaker worked in silence: the recordkeeper. They were soldiers temporarily reassigned, chaplains armed with nothing but a prayer book and a notebook, nurses with pencils tucked into apron pockets, and civilian clerks whose job was not to save the living, but to ensure the dead were not forgotten.

They understood that identity was fragile. That a name, once lost, might never be recovered. And that every scrap of record, each scrawled list, torn letter, or label stitched into a collar, was a lifeline to someone's grief.

From these quiet, imperfect acts of documentation came one of the most profound legacies of the Civil War: the first national archive of America's military dead. The Quartermaster General's Office, initially overwhelmed by logistical demands, began to standardize burial efforts by 1864. Field clerks and inspectors were assigned to mark graves, recover personal effects, and attempt identification. Their tools were simple: slates, chalk, waxed ledgers, and folding desks carried on wagons.[18] They often walked battlefields days or weeks after fighting ended, faces covered against the stench, hands blackened by wet ink and mud.

One Union chaplain, William Earnshaw, described crouching beside makeshift graves as his ink froze and the names blurred. *"We buried as many by prayer as by name,"* he wrote. Others used charcoal, bayonet tips, even blood when no pen could be found. Each name saved was a small defiance against oblivion.

Mistakes were inevitable. Names were misspelled, regiments were confused, and records were damaged by rain, rats, or careless handling. But still, they tried.[19]

For every name recorded, a family was spared the agony of unknowing.

For every grave marked, a future question could find an answer.

This was the emergence of what historian Drew Gilpin Faust calls "administrative mourning," a newly institutionalized way of managing grief.[20] In a war where thousands fell far from home and many were never recovered, recordkeeping became not just a bureaucratic task, but a moral mission.

It declared:

You were here.

You mattered.

You will not disappear.

Many of these recordkeepers were clergy. Army chaplains like Father William Corby, who served with the Irish Brigade, often performed last rites, buried the fallen, and noted their names and final words in personal journals.[21] These notes were sometimes later copied into unit rolls or shared with grieving families. Civilian pastors also served in field hospitals and camps, writing letters home for the dying and maintaining rudimentary burial logs when official systems failed.

Black and white nurses also served this role. In makeshift hospitals or behind the lines, they sometimes became the last witness to a soldier's life. They remembered details when paperwork did not. A favorite song, a home state, a name muttered through fever.[22]

After the war, this patchwork of documentation expanded into a more formalized national effort. The Freedmen's Bureau compiled burial and refugee records across the South, especially for the formerly enslaved and Black Union troops. The Ladies' Memorial Associations kept regional death rolls and grave maps, particularly for Confederate dead.[23]

Perhaps the most famous postwar recordkeeping initiative came from Clara Barton and her Office of Missing Soldiers. Between 1865 and 1868, Barton and her staff processed more the 63,000 inquiries and helped identity or locate over 22,000 soldiers.[24] Letters poured in from every corner of the country, each one a plea:

Was he found?

Where is he buried?

Can you send me something—anything?

Sometimes the answer was a grave location. Sometimes a note scribbled in the margin: *unknown, but buried with care.* Sometimes the news was devastating. But it was news—an anchor in the storm of unknowing.

The act of recordkeeping became an unexpected form of restoration. One name recovered could tether a widow's grief. One properly marked grave could halt generations of uncertainty. In a time when battlefield deaths often meant disappearance, to write a name down was to defy the void.

This was a different kind of monument, not built in stone, but in ink.

And long before the war's statues rose, before speeches were made or wreath laid, there were people—tired, ink-stained, and mostly forgotten—who tried to save what they could.

They could not stop the dying. But they could try to name the dead. And in doing so, they gave memory to a body to hold onto.

The clerks who transcribed names into ledgers worked in relative safety, far from the sounds of battle. But some witnesses to death served closer to the blood—women who nursed the dying, prepared the dead, and refused to let their sacrifice go unrecorded.

Among them was a woman who had been born enslaved, who had fled to freedom, and who would become one of the war's most remarkable—and least remembered—chroniclers.

Spotlight: Clara Barton and the Office of Missing Soldiers

Clara Barton's postwar work was not simply humanitarian—it was revolutionary. She founded the Office of Missing Soldiers in 1865 and operated it from a cramped room in Washington, D.C., with only a few volunteers. Over the next three years, they answered more than 63,000 letters and helped identify 22,000 missing men.

Barton did what no government agency had yet managed: she gave names back to the forgotten. She called this *"bringing them home in memory."*

Her office walls were lined with pinned letters, hand-drawn maps, and photos of the lost. This was America's first national mourning center—not in marble, but in paper and hope.

The work would follow her across an ocean. In 1869, Barton traveled to Europe and encountered the International Red Cross, founded just five years earlier on principles she recognized: impartial care, systematic documentation, the insistence that suffering demanded organized response. She saw in Geneva what she had built in Washington and understood that the Office of Missing Soldiers had been, without knowing it, a rehearsal.

In 1881, she founded the American Red Cross. The tracing work she had pioneered during the Civil War—the letters, the lists, the refusal to let the missing remain unnamed—would become a global mission. The thread that began at Antietam would stretch to Solferino, to the trenches of France, to every conflict that followed.

The Rise of Embalming: Death Becomes an Industry

"For fifty dollars, I can send your boy home looking like he is only sleeping."—Battlefield embalmer, 1862

The Civil War did more than create national cemeteries—it created the American funeral industry.[25]

Before the war, embalming was a curiosity, practiced mainly by medical schools on cadavers for anatomical study.[26] The dead were buried quickly, usually within a day or two, because preservation was impossible. Families gathered around deathbeds, washed and dressed their own dead, and held wakes in parlors and kitchens.[27]

The war changed everything.

Suddenly, thousands of young men were dying hundreds of miles from home. Their families wanted them back—not as bones in a box shipped months later, but as recognizable sons who could be mourned properly.[28] The demand for preservation created an industry overnight.

Enterprising embalmers followed the armies. Dr. Thomas Holmes, often called the "father of American embalming," claimed to have personally preserved over 4,000 bodies during the war.[29] He and others set up tents near battlefields, offering their services to families who could afford them. For fees ranging from $25 to $100—a substantial sum—they promised to preserve and ship bodies home.[30]

The process was crude by modern standards. It involved arterial injection of various chemical solutions, including arsenic compounds, followed by careful packing in sealed coffins.[31] But it worked. Bodies could survive the journey home by rail, arriving in condition for open-casket funerals.[32]

The embalming of Colonel Elmer Ellsworth in 1861 brought the practice to national attention.[33] Ellsworth, a young officer and friend of Lincoln, was killed in Alexandria, Virginia. His embalmed body lay in state at the White House before his funeral, looking peaceful and almost alive. The newspapers marveled at the preservation.[34]

But the most famous embalming of the war was Lincoln himself. After his assassination in April 1865, the president's body was embalmed and carried by funeral train from Washington to Springfield, Illinois, a journey of nearly two weeks including stops in major cities where thousands filed past his open casket.[35]

The embalming held. Lincoln's final tour became both a national mourning ritual and an advertisement for the possibilities of prese rvation.[36] The war transformed death from a domestic event into a commercial service. Embalmers became undertakers became funeral directors.[37] The parlor wake moved to the funeral home. And the dead, increasingly, were prepared by professionals rather than family.

This was not merely a practical shift. It was a spiritual one. Death became something to be managed, outsourced, professionalized. The intimate act of caring for one's own dead began to fade from American life, replaced by industries that promised to handle what families could no longer bear to do themselves.

Photography and the Visible Dead

"For the first time, the dead stared back."

In October 1862, Mathew Brady opened an exhibition in his New York gallery that changed how Americans understood war.[38] The photographs, taken at Antietam by his assistants, Alexander Gardner and James Gibson, showed something the public had never seen: actual corpses on an actual battlefield.

The images were shocking. Bodies lay bloated and twisted in fields and fence rows. Faces—when visible—bore expressions of agony or surprise. There was no heroism in these photographs, no glory. Only death in its raw, unvarnished reality.[39]

"Mr. Brady has done something to bring home to us the terrible reality and earnestness of war," wrote *The New York Times*. "If he has not brought bodies and laid them in our dooryards and along the streets, he has done something very like it."[40]

The exhibition drew enormous crowds. People who had read about casualties as statistics now saw them as sons, brothers, husbands. The dead were no longer abstractions. They had faces.

This visual confrontation with death had profound effects on the demand for proper burial. Families who saw the photographs—the neglected corpses, the hasty graves—insisted that their own dead be treated differently. The pressure for national cemeteries, identification systems, and dignified burial grew in part because photography had made the alternative visible and unbearable.

Photography also served identification purposes. Some families sent portraits to the front, asking soldiers to wear them into battle so that their bodies might be recognized. Others hired photographers to capture the faces of the dead before burial, creating records that might later help identify the missing.

The camera became a tool of mourning. Post-mortem photography—portraits of the dead, often posed as if sleeping—became common during the war years. Families who could not be present at a soldier's death commissioned photographs of the body, creating visual memorials to replace the deathbed vigils they had been denied.

The Civil War was the first photographed American war.[41] And in making death visible, it made the demand for dignified treatment of the dead impossible to ignore.

Artifact: A Wooden Grave Marker from Gettysburg

Recovered during 20th-century restoration work at Gettysburg National Cemetery, the weathered wooden board bore a faint hand-carved name: "Lt. R. M. Jenkins, VA." Below it: *"Died brave. Buried in haste. Remember him."*

Before marble and regulation, remembrance lived in wood, in hands that carved under torchlight, in the fleeting attempt to resist the anonymity of war.

Such artifacts remind us that for every engraved headstone we now walk past, there was once a marker made from fencepost and grief.

Part C: The Caretakers-The Women Who Carried the Grief

"We laid them to rest with our own hands. Not because we were told to—but because no one else would."—Civilian

volunteer, Tennessee, 1863.

While the battlefield belonged to men, the mourning belonged to women.

The Civil War tore through towns and farmland, across fences and front porches. It entered kitchens and sanctuaries. And when the smoke cleared, it was often women—mothers, daughters, wives, freedwomen, and farmhands—who emerged from cellars and chapels to face what had been left behind: the dead.

In the absence of military burial details, identification systems, or national protocols, women stepped forward. They buried bodies abandoned on fields, behind barns, and near rail lines. They dug with garden tools, carried water in buckets, and fashioned grave markers out of wood scraps and linens.[42] Some buried neighbors. Others buried strangers. All buried history.

Their weapons were not muskets, but cloth and grief.

In towns like Winchester, Virginia, and Franklin, Tennessee, civilian women ventured onto battlefields the morning after cannon fire had ceased, combing through the bloodied grass for the fallen.[43] They washed bodies with well water, stitched shrouds from repurposed curtains, and sang hymns over shallow graves. In the North and South alike, women transformed yards, orchards, and churchyards into impromptu cemeteries. Someone had to.

Some of these women were widows or sisters searching for loved ones. Others had no personal connection to the fallen. Still, they bore witness. They wrote down names if found, folded coats with letters still in the pockets, and preserved locks of hair or buttons in the hope that one day, someone might come asking.

This work was emotional, yes, but it was also deeply political.

To bury the dead was to make a statement: that these men mattered. That their lives had weight, even if the war had sought to grind them into silence.[44] Every act of care—laying a cloth across a face, tying a name to a wooden board, writing to a far-off family—was an act of resistance against the dehumanizing machinery of war.

In the Union North, women like Clara Barton rose to national prominence. After nursing wounded soldiers in battlefield hospitals, Barton recognized that the grief of the war was not ending—it was dispersing into thousands of households desperate for word. She founded the Office of Missing Soldiers, through which she and a handful of volunteers answered more than 63,000 letters and helped identify over 22,000 missing Union dead.[45]

But Barton was not alone. Thousands of unnamed women served as caretakers of memory, white and Black, enslaved and free. In the South, where Confederate dead were often excluded from federal cemetery efforts, women organized themselves into Ladies' Memorial Associations.[46] These groups raised funds, located graves, reinterred soldiers, and lobbied for monuments, shaping the literal and ideological landscape of Southern mourning. It is essential to acknowledge that many of these Southern women operated within the ideology of the Lost Cause, seeking not only to honor the dead but to reframe the narrative of the war itself.[47] Yet even within this context, their work highlights that women did not simply mourn; they organized, strategized, and curated national memory.

Clara Barton's counterpart in organized care was Dorothea Dix, appointed Superintendent of Army Nurses in 1861, the first woman to hold a major federal appointment during wartime. Dix had spent decades reforming mental asylums and prisons. Now, she turned her formidable will toward the chaos of military hospitals. She recruited, trained, and deployed over 3,000 women nurses, establishing standards for care that had never existed in American warfare. Her nurses did not only tend to the living, they sat with the dying, recorded final words, and wrote the letters that told families their sons would not return. Dix herself was exacting, difficult, and relentless. The work required nothing less. And though her nurses served in hospitals rather than on burial details, their labor was inseparable from the larger project of wartime care: ensuring that death, when it came, was witnessed, documents, and mourned.

Black women, often ignored in official records, also played a crucial role. Enslaved women were conscripted to bury bodies after battles swept across plantations. After emancipation, freedwomen became caretakers in national cemeteries, sometimes hired to tend graves or clean the grounds.[48] Others volunteered, returning to burial sites to plant flowers, repair markers, or sit vigil on days of remembrance.

Their stories are largely invisible in public commemorations. But they exist in faded letters, oral traditions, and the quiet persistence of generations who continued to visit, clean, and care for the graves long after others forgot. Some women wrote journals or memoirs. Others preserved battlefield diaries or pressed flowers from a soldier's grave between the pages of a Bible. Their mourning passed into lullabies, family stories, church rituals, and national traditions like Decoration Day, which would later become Memorial Day.[49]

They remembered when no one else did.

Their labor was unpaid. Their names were rarely etched in stone. But they were, in every sense, caretakers of national memory.

To care for the war dead was to claim a voice in the aftermath. To lay a body to rest was to resist forgetting. To grieve openly was to politicize pain in society unsure how to reckon with its own destruction.

These women teach us that war does not end at the signing of peace.

It ends when someone kneels down, places a hand on the earth, and says, "You will not be lost."

Interlude: The Ones Who Dug in Silence

"They told us to dig. So we dug. But we did not forget what we were doing."

In Virginia's clay, in Georgia's red earth, in Tennessee's muddy hills, the ground opened for the dead—dug not by soldiers or generals, but by hands newly freed.

These men had carried cotton and cane. Now they carried stretchers and spades.Some dug out of duty. Others because it was all that was offered. But many sang as they worked. Whispered prayers. Made sure names were not lost, even if the world wouldn't write them down.

They laid the bodies in the ground, yes.But they also buried the illusion that freedom could wait.

The men who dug in silence were not nameless by accident. They were nameless by design, rendered invisible by a system that had long refused to see them as fully human. But to understand the full scope of their labor, we must trace their journey from bondage to battlefield, from property to caretaker, from silence to sacred witness.

They came by the thousands, fleeing toward Union lines with nothing but the clothes on their backs and the hope of freedom in their hearts. The army that received them had a name for them—*contraband*—and a use for them that few white soldiers would accept.

They would bury the dead.

Contraband and Caretaker: The Burial Labor of the Newly Free

"They called us 'contraband' – property seized from the enemy. But we were the ones who buried their dead with dignity."—Testimony recorded by the Freedmen and Southern Society Project

When enslaved people fled to Union lines, they were not immediately recognized as free. Instead, they occupied a legal limbo—classified as "contraband of war," enemy property subject to confiscation. The term was dehumanizing, yet it also opened a door. As contraband, these men and women could remain behind Union lines. And there, in the shadow of the army, they became indispensable.

The Union Army discovered early in the war that it lacked sufficient manpower for the grueling labor that sustained military operations: digging fortifications, building roads, loading supplies, and—inevitably—burying the dead. White soldiers resisted such work when possible, viewing it as beneath the dignity of fighting men. The solution was found in the contraband camps that swelled around Union positions throughout the occupied South.[50]

By 1862, tens of thousands of formerly enslaved people had gathered in camps like Fortress Monroe in Virginia, at Cairo in Illinois, at Corinth in Mississippi, and dozens of others. Conditions were often desperate, the camps plagued by overcrowding, disease, inadequate food and shelter. Yet the camps also became sites of remarkable self-organization, as freedpeople established schools, churches, and mutual aid networks.[51]

The Union Army, recognizing the labor resource these camps represented, began systematically employing contraband workers. They were paid—though at rates far below those of white soldiers—and assigned to the most arduous and unpleasant tasks. Among the most demanding was burial duty.[52]

The scale of death in the Civil War overwhelmed all existing systems. After major engagements, bodies lay in fields for days or weeks. The stench of decomposition spread for miles. Disease threatened the living. Someone had to collect, identify, and inter the dead—and increasingly, that someone was Black.

Contraband burial details operated at nearly every major battlefield in the Eastern and Western theaters. At Antietam, where over 23,000 men fell in a single day, local freedpeople joined the burial crews that worked for weeks to inter the dead.[53] At Shiloh, contraband laborers dug mass graves for Confederate dead while Union soldiers received individual interment—a hierarchy of death that mirrored the hierarchy of the living.[54]

The work was physically devastating. Men dug in mud and blood, often without gloves or adequate tools. They handled bodies in every state of decomposition, from the newly fallen to those weeks dead. The summer heat of Southern battlefields accelerated decay, while the winter cold froze limbs into positions that made burial nearly impossible. There were no masks, no sanitation protocols, no psychological resources.[55]

Yet the testimony that survives suggests something more than mere labor. Freedpeople brought to this work a spiritual understanding of death and burial rooted in African and African American traditions.

The dead, in many West African cosmologies, remained connected to the living, and improper burial could disrupt that connection, leaving spirits restless and communities unprotected.[56] To bury the dead—even enemy dead, even the bodies of men who had fought to preserve slavery—was a sacred act.

Witnesses recorded freedmen singing as they worked. Spirituals rose from the burial pits—songs of deliverance, of crossing over, of rest:

"Swing low, sweet chariot, Coming for to carry me home..."

These were not performances for white observers. They were rituals of passage, offered to the dead by people who understood what it meant to be denied dignity in life and who refused to let death perpetuate that denial.[57]

Some contraband workers took pains to preserve what identity they could. When dog tags or letters were found, they were set aside. Names were spoken aloud before bodies were lowered into graves. One freedman, interviewed decades later by a WPA worker, recalled: "We didn't know them boys. But we knew somebody loved them. So we sang for them same as we'd want someone to sing for us."[58]

The work extracted a terrible toll. Typhoid, dysentery, and infections from contact with decaying flesh swept through burial details. Many contraband laborers sickened and died performing work the army had deemed too dangerous or degrading for white soldiers. Their deaths were rarely recorded with the same care as those of the men they buried. They became, in death, as invisible as they had been in life.[59]

And yet, their labor built the infrastructure of American military memory. The national cemeteries that arose from the Civil War (Antietam, Gettysburg, Arlington) were dug, in large measure, by Black hands. The orderly rows of white headstones that visitors walk today were made possible by men whose own graves, more often than not, bear no name at all.

To remember the Civil War dead, we must also remember those

who buried them.

They were not soldiers. They were not honored with medals or monuments. But they carried the weight of a nation's grief, and they carried it with grace.

The contraband laborers who buried the Union dead were civilians—refugees, not soldiers. But as the war progressed and the Lincoln administration authorized the enlistment of Black men, a new category of burial worker emerged: the soldier who fought beside the men he would later inter.

For the United States Colored Troops, burial duty carried a weight unlike any their white counterparts experienced. They were burying not only comrades but the cause of freedom itself.

From Contraband to Soldier: The Continuum of Black Burial Labor

"First they told us to dig. Then they told us to fight. In the end, we did both."—Attributed to a veteran of the 54th Massachusetts

The distinction between contraband laborer and soldier was, for many Black men, a matter of timing rather than transformation. The same hands that had dug graves as civilians in 1862 might hold rifles as soldiers in 1864—and then return to burial duty after battle.

The Union Army's reliance on Black labor for burial work did not diminish with the creation of the USCT. If anything, it intensified. Black soldiers were disproportionately assigned to fatigue duty of all kinds, and burial details ranked among the most common assignments. After engagements where white and Black troops had fought together, it was often the Black regiments that remained behind to inter the dead.[60]

This was not coincidence. It reflected the persistent assumption shared by many white officers and enlisted men alike that Black soldiers were better suited for labor than for combat, despite mounting evidence to the contrary. The valor displayed at Fort Wagner, Milliken's Bend, and dozens of other engagements did not erase prejudice. It merely coexisted with it, uncomfortably, in the same army.[61]

For Black soldiers, burial duty carried meanings that their white counterparts could not fully comprehend. They buried men who had died for a cause inseparable from their own freedom. They buried comrades who had shared the particular dangers faced by Black troops—the threat of enslavement or execution if captured, the knowledge that Confederate forces often refused to treat them as legitimate prisoners of war. And they buried, sometimes, men who had been murdered rather than killed in battle, victims of atrocities that the laws of war should have prevented.[62]

The emotional and spiritual weight of this labor cannot be measured in official reports. It lives instead in fragments: oral traditions passed down through families, brief mentions in pension files and veterans' testimonies, songs that witnesses recalled hearing rise from burial details at work.

They were soldiers. They were laborers. They were witnesses.

And they were the foundation upon which American military memory would be built—even as that memory systematically excluded them.

The United States Colored Troops: Serving the Dead They Fought Beside

"We fought for our freedom. And then we buried our brothers."—Serg

eant William H. Carney, 54th Massachusetts Infantry

The establishment of the United States Colored Troops (USCT) in 1863 transformed the Union war effort. By war's end, nearly 180,000 Black men had served in the U.S. Army, and another 18,000 in the Navy. They fought in over 400 engagements, from minor skirmishes to major battles like Fort Wagner, the Crater, and Nashville. Approximately 40,000 died in service—the majority from disease, but thousands in combat.[63]

What is less remembered is that USCT soldiers also served extensively on burial details, often interring the very comrades beside whom they had charged into battle. The Union Army's approach to USCT burial duty reflected the racial assumptions of the era. Black soldiers were frequently assigned to labor details that white regiments avoided, and burial work ranked among the most dreaded. After engagements in which white and Black troops had fought together, it was often the Black soldiers who remained behind to collect and bury the dead of both races.[64]

The experience was psychologically complex. USCT soldiers were burying men who had fought for their freedom; but they were also burying men who had, in many cases, viewed them as inferior, who had opposed their enlistment, who had doubted their courage. They buried former slaveholders' sons. They buried men who had been raised to believe Black people were property, not persons.[65]

And they buried their own.

The burial of Black soldiers killed in action was often conducted with less ceremony and care than that afforded white troops. After the assault on Fort Wagner in July 1863, Confederate forces buried the dead of the 54th Massachusetts Infantry in a mass grave, reportedly declaring that Colonel Robert Gould Shaw, the regiment's white commander, would be "buried with his niggers" as an insult.[66] What the Confederates intended as degradation, the 54th's survivors—and abolitionists across the North—transformed into a symbol of honor. Shaw's family declined efforts to recover his body, stating that they could imagine "no holier place" for him to rest than beside the men he had led.[67]

But not all USCT dead received even this grim dignity. In the aftermath of the Fort Pillow massacre of April 1864, where Confederate forces under Nathan Bedford Forrest murdered Black soldiers attempting to surrender, the dead were left unburied for days. When Union forces returned to the site, they found bodies mutilated, some apparently buried alive. USCT survivors were tasked with the harrowing work of recovering and interring their murdered comrades.[68]

Such experiences left deep scars. The trauma of burying men killed not merely in battle but in atrocity—men who had been denied the protections of the laws of war—compounded the ordinary horrors of Civil War burial work. USCT soldiers knew that if captured, they faced enslavement or execution rather than imprisonment. To bury comrades who had suffered such fates was to confront the precariousness of their own survival.[69]

Yet USCT burial details also became spaces of resistance and memory-making. Soldiers ensured that their dead were buried with whatever ceremony circumstances allowed. Chaplains (the USCT included fourteen Black chaplains by war's end) conducted services that drew on the traditions of the Black Church. When official records failed to document Black deaths adequately, soldiers and chaplains kept their own lists, their own memories.[70]

After the war, Black veterans would be among the most dedicated participants in Decoration Day observances, honoring the dead with flowers, flags, and prayer. They understood, perhaps better than anyone, the cost of the war and the fragility of the freedom it had won.

Their service did not end when the guns fell silent.

They had buried freedom's martyrs. Now they would fight to ensure those martyrs were not forgotten.

The USCT had proven that Black men could fight and die for the Union with courage equal to any white regiment. But the nation that accepted their sacrifice was not yet prepared to honor it equally.

Even in death—especially in death—the color line endured.

Segregated in Death: The Unequal Graves

"They died for the same flag. They were not buried beneath the same honor."

The promise of the national cemetery—that all Union soldiers would be buried with equal dignity, in orderly rows without distinction of rank—was a democratic ideal. But ideals, in Civil-War America, had a way of bending to the color line.

Black soldiers who died in Union service were technically entitled to burial in national cemeteries alongside white comrades. In practice, the reality was more complicated. At many sites, USCT dead were interred in separate sections, their graves clustered at the edges of cemeteries or in areas designated—explicitly or implicitly—as "colored" sections.[71]

The segregation of military cemeteries reflected the broader ambivalence of the Union toward its Black soldiers. The USCT had proven their courage on battlefields across the South. They had stormed fortifications, held defensive lines, and died in numbers that silenced many—though not all—of the skeptics who had doubted their valor. Yet the army that accepted their sacrifice was not yet prepared to accept their equality.[72]

At Arlington National Cemetery, established on the confiscated estate of Robert E. Lee, segregation persisted into the 20th century. Freedmen's Village, a settlement of formerly enslaved people adjacent to the cemetery, had its own burial ground separate from the orderly rows of military graves. When the village was eventually demolished, many of its graves were lost or relocated without adequate records. The Black dead, even in the nation's most hallowed cemetery, remained apart.[73]

This disparity extended beyond geography to commemoration itself. White soldiers' graves were more likely to be marked with permanent headstones; Black soldiers' graves more often bore wooden markers that rotted within years. White soldiers were more likely to be identified and recorded; Black soldiers more often ended up in collective graves or listed simply as "unknown." When memorial associations erected monuments to Civil War dead, Black soldiers were rarely represented. When veterans' organizations gathered, Black veterans were often excluded or relegated to subordinate roles.[74]

The Freedmen's Bureau, which assumed responsibility for many aspects of Black life in the postwar South, also took on responsibility for Black military dead. Bureau agents worked to identify and properly inter USCT soldiers scattered across Southern battlefields and hospitals. But the Bureau was underfunded, understaffed, and, after 1872, abolished entirely. Much of its burial work remained incomplete.[75]

Some communities resisted the erasure. Black veterans' organizations, including local posts of the Grand Army of the Republic (where they were admitted) and independent Black veterans' groups (where they were not), maintained their own memorial traditions.

They decorated graves on Decoration Day, organized parades, and told stories that white America preferred to forget.[76]

Black churches became repositories of memory. Congregations preserved the names of men who had served, honored their widows and orphans, and integrated military sacrifice into the ongoing narrative of Black freedom. The church cemetery, even where national cemeteries were inaccessible or unwelcoming, became a space where Black soldiers could be buried with honor.[77]

These parallel traditions of memory—one official and increasingly white, one grassroots and persistently Black—would diverge throughout the late 19th and early 20th centuries. As the ideology of the "Lost Cause" romanticized the Confederacy, and as North and South reconciled around a shared commitment to white supremacy, Black contributions to Union victory were systematically minimized. The USCT faded from textbooks. Their cemeteries fell into neglect. Their descendants continued to remember, but the nation chose to forget.[78]

Only in recent decades has scholarship begun to recover what was lost. Historians have documented the extent of segregation in Civil War cemeteries and memorials. Activists have pressed for recognition of Black soldiers' contributions. In some cases, graves have been rededicated, monuments erected, and stories retold.

But the disparity in death remains a testament to an uncomfortable truth: that the nation Black soldiers died to save did not yet consider them fully part of itself. Their graves, set apart even in cemeteries built to honor shared sacrifice, stand as silent witnesses to the long struggle for equality that the war had begun but not completed.

It would take more than a century for serious efforts at equity to begin—for historians to document the segregated sections, for advocates to push for recognition, for the full story of Black burial labor to enter public memory.

The graves are still there, waiting to be seen as they deserved to be seen from the beginning: as testaments to courage, sacrifice, and a nation that did not yet know how to honor all its dead equally.

Spotlight: Susie King Taylor – Nurse, Teacher, Witness

She was born enslaved in Georgia in 1848. She died free in Boston in 1912. During her life, Susie King Taylor became one of the most remarkable—and least remembered—witnesses to the Civil War and its aftermath.

At age fourteen, Taylor escaped to Union lines and attached herself to the 33rd United States Colored Troops (originally the 1st South Carolina Volunteers), one of the first Black regiments organized during the war. Though officially employed as a laundress, Taylor served in capacities far beyond that title. She nursed the sick and wounded. She taught soldiers to read and write—a skill forbidden to them under slavery. And when battles ended, she helped prepare the dead for burial.[79]

In her memoir, *Reminiscences of My Life in Camp* (1902)—the only published account of the war by a Black woman who served with Union troops—Taylor described the full spectrum of military deathcare. She cleaned wounds that could not be healed. She sat with dying men and recorded their final words. She assisted in washing bodies before interment, a task that combined medical necessity with spiritual care.[80]

"I assisted in caring for the sick and injured comrades," she wrote, "and was often called upon to help with the dead."[81]

Taylor understood that the work she performed—nursing, teaching, burying—was not ancillary to the war effort but central to it. The Black soldiers she served were fighting not merely for Union but for their own humanity. To care for them in sickness and death was to affirm that humanity against a system that had denied it.

After the war, Taylor remained dedicated to veterans' causes. She organized a Women's Relief Corps chapter in Boston and advocated for Black veterans' benefits at a time when such advocacy was neither popular nor safe. She lived to see the promises of Reconstruction betrayed, the rights of Black citizens stripped away, and the contributions of Black soldiers erased from national memory.

Her memoir was a protest against that erasure. "There are many people who do not know what some of the Colored Women did during the war," she wrote in her introduction. "I have made this book to tell you."[82]

Susie King Taylor was not merely a witness to the Civil War.

She was a caretaker of its memory. And she refused to let that memory be forgotten.

Taylor's memoir preserved what official records could not: the human texture of Black military service, the intimacy of care amid carnage, the sacred labor of tending the dead. But she was not alone in her determination to remember. Throughout the USCT, chaplains and officers kept their own records. They maintained unofficial lists of the fallen, not for the Army but for posterity. One such record survives from the most famous Black regiment of the war.

Artifact: The Burial Ledger of the 54th Massachusetts

Among the surviving records of the 54th Massachusetts Infantry—the first Black regiment raised in the North—is a small leather-bound ledger, now held in the Massachusetts Historical Society. Its pages contain the names of soldiers who died during the regiment's three years of service, from the assault on Fort Wagner in July 1863 to the final weeks of the war.

The ledger was maintained by the regiment's chaplain, Reverend Samuel Harrison, one of the first Black chaplains commissioned in the U.S. Army. Harrison understood that the official casualty reports would not suffice. The Army's bureaucracy, overwhelmed by death on an industrial scale, would not remember each man as he deserved to be remembered.[83]

So he kept his own record.

Each entry includes the soldier's name, company, date of death, and (where known) the cause and place of death. Some entries include additional notes: "Died of wounds received at Wagner." "Killed in the assault." "Died of fever at Morris Island."

A few entries are more personal: "A good soldier and a faithful man." "He was loved by all who knew him." "May he rest in peace."

The ledger was not official. It conferred no benefits, generated no pensions, triggered no notifications to next of kin. It was, in military terms, unnecessary.

But Harrison knew that memory is never unnecessary.

The men whose names he inscribed had either been enslaved, born to enslaved parents, or lived their entire lives under the shadow of slavery. Many had been told they could not fight, that they lacked the courage or discipline for military service. They had proven those doubts wrong on the ramparts of Fort Wagner, in the trenches of Charleston, in skirmishes and sieges across the South Carolina Low-country.

They deserved to be remembered.

And so Harrison wrote.

The ledger is not a military document. It is a memorial—one of the first to honor Black soldiers not as abstractions or symbols but as individuals with names, histories, and claims upon the future.

Each entry is an act of witness.

Each name is an act of love.

The ledger of the 54th Massachusetts was a protest against forgetting, a small act of witness in a war that threatened to swallow individual lives into mass anonymity. It was one of thousands of such acts—the name scratched into a wooden board, the letter folded into a pocket, the prayer whispered over a fresh grave.

Together, these acts of remembrance did more than honor the dead.

They made a nation.

PART D: THE LEGACY-THE FIRST DECORATION DAY

"Give me the man who bends his body to the work of freedom."—Frederick Douglass 1863.

The story of Memorial Day, as commonly told, begins in 1868, when General John A. Logan, commander of the Grand Army of the

Republic, proclaimed May 30 as a day for decorating the graves of Union soldiers. Logan's proclamation is well documented. What is less remembered—indeed, what was actively suppressed for generations—is that freedpeople in Charleston, South Carolina had already established the tradition three years earlier.[84]

In the spring of 1865, Charleston lay in ruins. The city that had inaugurated secession, that had fired on Fort Sumter, that had served as a symbol of planter aristocracy and Confederate defiance, was now occupied by Union forces—including regiments of United States Colored Troops. Among the USCT soldiers who marched through Charleston's streets were men who had been enslaved in the surrounding Lowcountry, returning as liberators.[85]

At the Washington Race Course, a horse-racing track northeast of the city, the Confederates had established a prisoner-of-war camp during the war's final year. Conditions had been appalling. Union prisoners, packed into the open-air infield without adequate shelter, food, or medical care, had died by the hundreds. Their bodies had been buried in a mass grave behind the grandstand, with no markers, no ceremony, no dignity.[86]

In the weeks following the Confederate surrender, Charleston's Black community—freedpeople, free Blacks, and newly arrived missionaries and teachers—organized to remedy this neglect. They exhumed the mass grave and reburied the Union dead in individual graves, each marked with a whitewashed wooden headboard. They built a fence around the new cemetery and erected an archway over the entrance bearing the words: "Martyrs of the Race Course."[87]

On May 1, 1865, more than 10,000 people gathered at the site to honor the dead. The crowd was predominantly Black, composed of freedpeople from Charleston and the surrounding plantations joined by white missionaries, teachers, and Union soldiers. Children from the city's newly established Black schools led the procession, carrying armfuls of flowers. Following them came Black women, members of the city's churches and mutual aid societies; then Black men; then Union troops, including the 54th Massachusetts and the 34th and 104th USCT regiments.[88]

They sang hymns. They listened to speeches. They prayed.

And then, rank by rank, they walked among the graves, laying flowers on each one.

It was, by all accounts, the first large-scale Decoration Day ceremony in American history.

The *New-York Tribune* reported on the event, as did several other Northern newspapers. But in the decades that followed, as the politics of Reconstruction gave way to the politics of reconciliation, the Charleston story faded from national memory. White Northerners and Southerners alike found it more comfortable to locate the origins of Memorial Day in white communities—Waterloo, New York, became the officially recognized birthplace in 1966—than to acknowledge that freedpeople had led the way.[89]

The erasure was not accidental. It reflected a broader pattern in Civil War memory that included the marginalization of Black agency, the romanticization of Confederate "valor," and the subordination of emancipation to reunion. To admit that freedpeople had originated Memorial Day was to admit that Black Americans had been central to the war's meaning, a truth that white America, North and South, was not yet ready to accept.[90]

The Charleston ceremony embodied everything this book has sought to recover: the sacred labor of burial, the insistence on dignity for the dead, the transformation of grief into collective memory. The freedpeople who reburied Union soldiers and decorated their graves understood that to honor the dead is to claim the future, that memory is power, and that those who control the rituals of remembrance shape the meaning of history.

They had no official sanction. They had no government funding. They had only their hands, their faith, and their determination to ensure that the men who had died for their freedom would not be forgotten.

On May 1, 1865, in a city still smoldering from war, formerly enslaved people invented Memorial Day. It is time we remembered.

The freedpeople of Charleston understood what military bureaucracies were only beginning to learn: that memory requires not only ritual but record. The graves they tended at the race course were marked. The names they could recover were spoken aloud. The ceremony they created was witnessed and reported.

But across the shattered South, thousands of other graves bore no names at all. The chaos of war had overwhelmed every system of identification. And in the absence of adequate records, countless families would spend the rest of their lives searching for sons, husbands, and fathers who had simply vanished into the earth.

The task of naming the dead—of giving grief an address—fell to those who held the pens.

The Cost of Living

"The war ended. Their work did not."

We know what burial labor demanded of the body. The records tell us that much. Fevers, infections, and dysentery swept through burial details like a second enemy. Men who dug graves in summer heat collapsed from exhaustion. Men who handled decomposing remains developed sores that would not heal. The physical toll was measurable, if rarely measured.

But what of the toll that left no scar?

The Civil War predates our modern vocabulary of trauma. There was no "PTSD," no "moral injury," no clinical language for what it meant to spend weeks lowering broken bodies into the ground. Soldiers called it "soldier's heart" or "nostalgia," vague terms for a suffering they could not name. For the burial workers, there was often no term at all.

Yet the evidence survives in fragments. In the pension files of veterans who could not sleep, who startled at sudden sounds, who told interviewers decades later that they still saw the faces. In the letters of nurses who described dreams filled with the dying. In the silence of freedmen who never spoke of what they had buried, even to their children.

One contraband laborer, interviewed in the 1930s by a WPA worker, said only: "I dug until I couldn't feel my hands no more. And then I kept digging. You don't forget that. You just learn to carry it."

They carried it home. They carried it into marriages, into churches, into the fields they plowed and the children they raised. Some found ways to speak; most did not. The nation that owed them recognition offered them silence instead.

The chaplains who whispered prayers over the nameless. The women who washed blood from shattered limbs. The clerks who transcribed name after name until the names lost meaning. The freedmen who sang spirituals into open graves. All of them paid a price that no ledger recorded.

We call this work "sacred labor." But sacred does not mean painless.

The caretakers of the Civil War dead did not merely bury bodies. They buried part of themselves as well.

Mourning Made a Nation

"The war ended. The mourning did not."

The Civil War did not just reshape the country's borders or break its political institutions. It reshaped how Americans understood death. It transformed how a nation mourned.

Before the war, grief had been a largely private affair. Death happened at home, surrounded by family, within reach of familiar rituals. The Civil War shattered that intimacy. It pulled fathers, sons, and brothers from hearth and community and scattered their bodies across hundreds of battlefields, miles away many never to return. In the way of such devastation, mourning became public. Collective. And increasingly, political.

The federal government, for the first time in its history, took responsibility for the dead—not just in burying them, but in remembering them. National cemeteries were founded. Clerical offices were opened. Ledgers were filled with names. And the collective act of grieving became embedded in the American civic experience.

To mourn, after the Civil War, was to assert that every life mattered, that every death deserved recognition, and that even in failure, the attempt to remember was a kind of redemption.

But this chapter is not only about the policies and systems that emerged. It is about the people who gave those systems meaning:

– The women who searched for sons and stitched grave markers from scraps

– The freedmen who dug graves with bare hands and sang over them in the dark

– The clerks who transcribed names into ledgers and called it sacred

– The chaplains and nurses who whispered blessings over the nameless

These were the first great rememberers in American war history.

They did not wear brass buttons or hold rank. They received no medals. But their labor—emotional, physical, spiritual—made mourning possible. They helped a fractured country begin to make sense of its dead. And in doing so, they helped it begin to piece itself back together.

The Civil War was the first American conflict in which the dead outnumbered the narratives that could hold them. It forced the country to create new languages of grief, ones written in ink and earth, in fabric and prayer.

Through their work, caretakers—seen and unseen—ensured that the war's silence did not swallow its memory. That the fallen were not reduced to numbers. That they were laid to rest with dignity, even if they had no name.

This mourning was not just for the sake of the dead.

It was for the sake of the living.

For the country's soul.

And still, at the heart of every national cemetery, of every roster, of every funeral flag folded with care, lies the same truth:

The dead matter.

But so do those who bury them.

By the war's end, the Union had buried over 300,000 men. Nearly half of them—more than 140,000—remained unidentified. Unknown.

Somewhere in Virginia, in Tennessee, in the Carolinas, a grave marker bears no name. A wooden board, weathered past reading. A stone that says only: "U.S. Soldier."

Somewhere, a woman folded her last letter and put it in a drawer. She never received an answer.

Somewhere, a freedman walked home from a burial detail, his clothes still smelling of the dead. No one asked what he had seen. No one recorded his name. He washed his hands in a creek, and the war, for him, was over—except that it wasn't. It never would be.

These were the caretakers who built America's memory of war.

They were not thanked. Many were not paid. Almost none were remembered. But the tools they forged—cemetery grids, identification protocols, recordkeeping—would not die with them. They would cross oceans, shaping the burial landscapes of France, Belgium, and every battlefield to follow.

The laws had been written. The labor had been done.

Now came the trenches.

Intermission

Between Wars: 1865-1917

The Tools of Grief, Unfinished

"The systems we build for the Civil War dead were not ready. They were never ready. And fifty years later, we would learn that lesson again."

The guns fell silent at Appomattox in April 1865. The nation buried its dead—or tried to. National cemeteries rose from cornfields and hillsides. Ledgers filled with names. And for a generation, Americans believed they had learned what war demanded of memory.

They were wrong.

The fifty years between the Civil War and World War I were not a pause in American military death. They were a slow forgetting, a period during which the systems born of grief calcified. The lessons grew distant, and the nation convinced itself that such slaughter could never happen again. When it did, America was no more prepared than it had been at Antietam.

This intermission traces the arc of that forgetting: the wars that tested but did not transform, the professions that emerged, the rituals that hardened, and the inequalities that deepened. It is the story of what happened between the wars. And why it wasn't enough.

The Forgotten War: Death in the Tropics (1898)

The Spanish–American War was supposedly glorious. It lasted barely four months. The battles—San Juan Hill, Manila Bay—became the stuff of legend, launching Theodore Roosevelt into the White House and America onto the world stage.

But the war's dead told a different story.

Of the approximately 2,900 American soldiers who died during the conflict, fewer than 400 fell in combat. The rest—over 2,500—died of disease: typhoid, malaria, yellow fever, dysentery.[1] They died in Cuban camps and Philippine jungles, in hospitals overwhelmed by illness, in transports where the sick lay packed together in holds that became floating morgues.

The military was unprepared. There were no adequate systems for identifying the dead in tropical conditions. Bodies decomposed rapidly in the heat and humidity. Embalming, which had become standard practice in the United States, was difficult or impossible in the field.[2] Many soldiers were buried where they fell, in graves that would be lost within months to jungle growth or monsoon rains.

For families at home, the experience was agonizing. Notification was slow and often inaccurate. The War Department, still operating with Civil War-era systems, struggled to track casualties across two oceans.[3] Letters went unanswered. Telegrams arrived weeks late. Some families learned of their sons' deaths only when comrades returned home with stories.

The war also marked America's first significant experience with overseas military cemeteries. In Cuba, the Philippines, and Puerto Rico, graves were scattered across battlefields, hospital grounds, and makeshift burial sites.[4] The question of what to do with these distant dead—whether to leave them where they lay or bring them home—would not be fully answered for decades.

The Spanish-American War should have been a warning. It demonstrated that modern warfare, even "splendid little wars," produced death on a scale that existing systems could not manage. It revealed the particular challenges of distant death: identification, preservation, communication, and the anguish of families separated from their grief by thousands of miles.

But the lessons were not learned. The war was too short, too triumphant, and too easily mythologized. The dead were folded into narratives of expansion and glory. And when the next war came—truly global, truly industrial—America would face the same problems again, only magnified beyond comprehension.

The Professionalization of Death

While the military slumbered, civilian deathcare transformed.

In the decades after the Civil War, embalming (once a wartime expedient practiced by itinerant operators) became a profession. State licensing boards began to emerge in the 1890s. Professional associations formed. Training schools opened, teaching the chemistry of preservation and the art of presentation.[5] The undertaker became the funeral director, and the trade became an industry.

By 1900, the National Funeral Directors Association counted

thousands of members across the country.[6] Embalming was standard practice for middle-class funerals. The funeral home—a dedicated space for viewing, mourning, and ceremony—had replaced the family parlor as the site of death ritual.[7] Americans increasingly outsourced their grief to professionals.

This professionalization had profound implications for military burial. It created a civilian expertise that the military lacked. When World War I began, the Army recruited funeral directors into the Graves Registration Service, drawing on skills developed in peacetime.[8] Embalming techniques refined in American funeral homes were adapted for battlefield conditions. The paperwork of civilian death—certificates, records, chains of custody—influenced military documentation.

But professionalization also widened the gap between the military's ad hoc systems and the standards Americans expected. Families who had grown accustomed to viewings and ceremonies found military burial—hurried, distant, often without a body—almost incomprehensible. The war would force a confrontation between civilian expectations and military realities that persists to this day.

Memorial Day and the Rituals of Remembrance

The work of memory did not stop with burial. In the decades after the Civil War, Americans built an elaborate culture of commemoration—one that would shape how the nation mourned all its future dead.

Memorial Day, born in the immediate aftermath of the war, became a national ritual. By the 1880s, the holiday was observed across the North and, in modified form, across the South.[9] Veterans marched. Speeches were delivered. And in cemeteries from Maine to California, citizens placed flowers on the graves of the fallen.

The ritual was democratic in aspiration. Every grave was to be decorated, every soldier remembered regardless of rank or origin. But in practice, commemoration reflected the inequalities of the living. Black veterans, who had participated in the earliest Memorial Day observances, were increasingly marginalized as the century wore on.[10]

Confederate memorialization, organized by women's groups and veterans' associations, constructed a mythology of the "Lost Cause" that erased slavery as the war's central issue and celebrated treason as heroism.[11]

Veterans' organizations shaped this commemorative landscape. The Grand Army of the Republic (GAR), the largest Union veterans' organization, lobbied for pensions, built soldiers' homes, and organized Memorial Day observances. At its peak in the 1890s, the GAR counted over 400,000 members.[12] Confederate veterans' groups, though smaller, wielded outsized influence in the South, erecting monuments and controlling the narrative of the war.

These organizations also served a quieter function: they helped veterans process what they had experienced. In an era before trauma was medicalized, the camaraderie of fellow soldiers provided a form of support that no other institution offered.[13] Men who could not speak of the war to their families found understanding among those who had been there. The veterans' encampment, the memorial parade, and the cemetery visit became rituals of acknowledgment, if not healing.

By the time these veterans began to die—by the thousands in the early 20th century—their organizations had institutionalized memory in ways that would outlast them. The American Legion, founded after

World War I, would inherit their mission and their methods.[14] The rituals of Memorial Day, the expectations of military burial, and the demand for official recognition all descended from the culture these Civil War veterans had created.

Jim Crow and the Segregation of Memory

The systems of unequal burial established during the Civil War did not fade with Reconstruction. They hardened.

As Jim Crow laws spread across the South and de facto segregation defined the North, the color line extended into death. Black veterans were buried in segregated sections of national cemeteries, if they were buried there at all.[15] Black casualties from the Spanish–American War and the Philippine–American War faced the same disparities their fathers had: delayed notifications, incomplete records, forgotten graves.

When Black soldiers did receive burial in national cemeteries, their graves were often clustered at the edges, in sections designated by race rather than regiment. The democratic ideal of equal rows beneath identical headstones—the promise of the national cemetery—was compromised from the start.[16]

This segregation was not merely spatial; it was archival. Records of Black soldiers were less carefully kept, more easily lost. Pension files of Black veterans were more frequently denied or delayed.[17] The commemorative culture that honored white veterans systematically excluded their Black comrades.

And yet, Black communities maintained their own traditions of remembrance. Black churches decorated graves. Black veterans' organizations—often excluded from the GAR—formed their own posts and held their own observances. The first Memorial Day, after all, had been created by freedpeople in Charleston. That tradition of communal memory, forged in the shadow of slavery and sustained through the decades of Jim Crow, would endure.[18]

When World War I came, it would call on Black men to serve—and die for—a nation that refused to bury them equally. The patterns of 1865 would repeat in 1918. And the struggle for equal recognition in death would continue for another century.

The Persistence of Unpreparedness

By 1914, the U.S. Army was a small, professional force of fewer than 100,000 men, scattered across coastal fortifications and posts.[19] It had no significant experience with mass mobilization since the Civil War. Its quartermaster corps, responsible for supply and logistics, had no dedicated capacity for handling large numbers of dead.

The lessons of the Civil War had not been institutionalized. There was no permanent Graves Registration Service. No standardized identification system. No established procedure for overseas burial or repatriation. The ad hoc arrangements of 1861–1865 had been dismantled, and nothing systematic had replaced them.

When the United States entered World War I in April 1917, it had to rebuild from scratch. The GRS was created as an emergency measure. Dog tags—two per soldier, one to stay with the body—were mandated.[20] Burial forms were designed, printed, and distributed. Cemetery sites were surveyed, and repatriation policies were debated.

All of this happened under fire, in the mud, as bodies accumulated faster than any system could process.

The fifty years between wars had not been wasted. The professionalization of deathcare, rituals of commemoration, and veterans' culture of mutual support all proved useful. But the fundamental problem remained: America had never developed a permanent infrastructure for military death. It responded to each war as if surprised by the dying.

That surprise cost thousands of families the closure they deserved. It left tens of thousands of soldiers in unmarked graves. It burdened the men assigned to burial duty with work for which they had no training and no support.

The Civil War had taught the nation that industrial slaughter demanded industrial memory. The nation had learned the lesson—and then, slowly, forgotten it.

Across the Atlantic, the guns of August began to speak.

And America's dead were waiting.

This intermission bridges the gap between the systems born in the Civil War and the systems tested—and often broken—in World War I. The tools of grief, forged in one conflict, proved inadequate for the next. And the racial, economic, and institutional inequalities of American life followed the dead into the ground.

The next chapter enters the trenches.

Reflection

My great-grandfather survived the trenches. He came home blind. The records don't specify exactly what took his sight. Gas, perhaps, or a blast injury, or an infection in a field hospital where the wounded outnumbered the supplies. What the records do contain are his memoirs: handwritten pages describing life in the trenches, dictated to my great-grandmother after his eyes failed him. He wrote about the mud. The rats. The constant shelling. And the bodies—everywhere, the bodies. Men who had been alive that morning were dead by afternoon, lying in the trenches alongside the living because there was no way to remove them, no pause in the fighting long enough to bury them properly.

They ate beside the dead. They slept beside them. They stepped over them and tried not to look.

Reading those pages, I thought about Ernest Hemingway—one of my favorite authors—who served as a stretcher-bearer on the Italian front. I had always known he carried trauma from the war, but I hadn't fully understood what being a "stretcher-bearer" meant. It meant carrying the wounded, the dying, and the dead. It meant witnessing death at close range, again and again, until the images burned themselves into memory. Hemingway's spare prose, his obsession with death, and his characters who cannot escape what they have seen—all of it makes more sense when you understand what he carried home from those trenches.

My great-grandfather carried it too. He lived with it for decades, blind but still seeing the things he had witnessed as a young man in France. This chapter is for him. And for all the young men who went into those trenches whole and came out broken—whether or not the wounds were visible.

Chapter Four

World War I: The Doughboys and the Dirt

"They weren't surgeons or generals. Most had never seen a dead body before the war. Some had barely left their hometowns. But there, amid the barbed wire and blood of Europe, they became the keepers of grief."

The First Global Burial CrisisWorld War I shattered not only borders and empires—it shattered the ways in which the world remembered its dead.

For the first time in history, nations faced mass death on an industrial scale. The Great War transformed Europe into a graveyard. Millions of soldiers perished in trenches, forests, and shell-based fields. Entire units were lost in a single day, their bodies buried in mud or never recovered at all. And amid this catastrophe, the burden of remembrance fell not to generals, but to ordinary men—many barely out of boyhood—who were ordered to stay behind after the fighting stopped, shovels in hand.

This chapter is not about strategy or diplomacy. It is about what happened after.

After the charge.

After the silence.

After the smoke.

It is about the ones who bent down in ruined fields to reclaim what war had left behind. The ones who carried bodies out of cratered woods, who tied name tags to mangled limbs, who wrote in ledgers with shaking hands. The ones who were tasked with turning death into memory.

It is the story of the Graves Registration Service, a military unit that grew out of necessity and grief. Formed during America's late entry into the war, the GRS became the quiet backbone of battlefield recovery and military burial efforts. Its members, often chosen with little explanation or preparation, became the stewards of the dead.[1]

But they were not alone. Across the war's shattered landscapes, caretakers emerged. Red Cross nurses logged names and folded keepsakes. Clergy wrote letters to mothers. Black laborers who buried the fallen by the hundreds. Villagers marked graves with stones, twigs, and prayer.

Each was part of a new, improvised system of remembrance—one forged in the mud and fear of the Western Front.

The Civil War had taught America the moral cost of anonymity in death. It had taught the value of names, of records, of national cemeteries. But World War I raised the stakes. It internationalized the dead, scattering them across oceans and continents. And it demanded a new answer to an old question: *how do we care for the dead when there are too many to count?*

This chapter follows that question. It walks beside the men who gathered the pieces. It listens to the clerks who tried to name the lost. It honors the silent rituals that bloomed amid the filth of war—rituals of faith, memory, and quiet defiance.

It also confronts the erasures:

The racial segregation of burial duty.

The families who never received notice.

The names that were never carved into stone.

Because remembrance, in war, is never equal. Neither is grief.

But still, across mud and barbed wire, they remembered. They buried, they wrote, they tried.

World War I did not only redefine war; it redefined what it meant to bury, to mourn, and to remember on behalf of a nation. And though the dead of that war now lie in clean rows beneath stone, it began with hands in the dirt. With men who were not trained for reverence, but who gave it anyway.

They were the gravediggers of the Great War.

And this is their chapter.

Interlude: Profile of a GRS Soldier (Based on Real Sources)

Private James F. Hanley of Pennsylvania arrived in France in May 1918 expecting to serve as a field medic. Instead, he was reassigned to a Graves Registration detail two weeks after arriving in the Argonne sector. In his wartime journal, he wrote:

"I keep a list of the names I've marked. Not because I'm told to, but because I'm scared I'll forget one. They start to blur after a while—faces, boots, what's left of a coat. One man had a wedding ring tied around his neck with string. I used it to ID him. That's the only time I've cried."

Baptism by Mud: The Call to Bury

"I thought I came to fight. They handed me a shovel."—Private, American Expeditionary Forces, 1918.

They weren't trained for this.

When young American soldiers stepped off troopships in France in 1917 and 1918, most carried visions of valor. The posters had promises adventure, and the recruiters had praised duty. But some men, barely eighteen and fresh from farms, factories, or cities they'd never left before, were pulled aside and given a different assignment.

They were to bury the dead.

There was no glory in the posting. It wasn't written on enlistment forms or praised in parades. Sometimes the orders came without explanation. A sergeant pointed, a name was read, and a shovel, not a rifle, was placed in the young man's hand.[2]

These men, suddenly reassigned to Graves Registration Service (GRS) duty or impromptu burial details, often had no medical or clerical training. Maybe they had steady hands, maybe they were calm, or maybe they just happened to be standing nearby when the need arose. It didn't matter. The work had to be done.

And so, their baptism into war wasn't gunfire.

It was blood.

It was bone.

It was mud.

The Western Front was unlike anything Americans had ever experienced. Artillery barrages lasted for days. Shellfire didn't just kill—it obliterated. Gas attacks melted faces and lungs. Machine guns cut bodies apart in seconds. Entire trenches collapsed under the weight of death and rain.[3]

In this chaos, recovery was a nightmare. Bodies decomposed in craters filled with water and blood. Corpses froze in awkward positions or melted into the muck. Uniforms were torn away, faces unrecognizable. The task of burial was not merely unpleasant. It was psychologically harrowing.

Graves Registration men worked in these conditions daily.

They ventured out between battles, often under sniper fire or in gas masks, armed only with stretchers, body bags, and rudimentary tagging materials.[4] Their mission was twofold: recover the body and identify the man.

Many wrote later that nothing in their lives had prepared them for what they had seen.

"I was just a farm kid from Iowa," one soldier recalled. "First week out there, I helped pick up what was left of a dozen men. That night, I puked until I passed out. The captain said that meant I was normal."[5]

They quickly learned how to dissociate, how to work in silence, how to stomach the stench of decay while noting a name on a ledger. They became intimate with death's textures—its weight, its color, its resistance in mud. They handled body parts like puzzle pieces, sometimes reassembling the dead so that a mother might receive a closed casket. They were taught to do their work with precision, even as everything around them collapsed.

And still, the work never became routine.

Every body was a story interrupted. A promise unkept. A life imagined, now broken in pieces. GRS personnel often knew they were the last to see these men whole. The last to touch them. The last to witness their passage.

Some whispered prayers. Others remained silent.

But in that silence was reverence.

The terrain added another layer of challenge. Trenches caved in, and the craters filled with rainwater to swallow the dead. Sometimes recovery had to wait for days or weeks, and by then, the remains were barely remains at all. "Sometimes," a GRS soldier write, "you didn't bury a man, you buried what you hoped was him."[6]

Yet even in this horror, these men found purpose. They cleaned boots before sending them home. They folded letters found on bodies. They traced names into notebooks with care. It was hard, thankless labor. But it mattered.

The job created something rare in war: a moment of stillness.

Of ritual.

Of witness.

They were not trained priests or physicians. They were not generals or surgeons. But they were caretakers of dignity.

And though they were forgotten in official reports, often absent from victory celebrations, they carried the weight of the war in the most human way possible: *one broken body at a time.*

Interlude: The Button

"All we had was a coat button. But it was his."

We found him at the edge of a flooded trench, half-submerged in mud, one boot gone, face obscured by weeks of rain and rot. No tags. No letters. Just a shredded coat pinned to what was left of his shoulder.

We almost didn't check the pocket, but I had a feeling. Don't know why. Inside, sewn behind the lining, was a single brass button. A clean one. Too clean. It didn't match the rest. I pulled it out and saw what someone—maybe his mother—had etched onto the back:

P. Roberts – 18th Inf. – Ill.

That was all. A name, a regiment, a place. But it was enough.

The paperwork came later. The card, the number, the grave marker. But that button was the first page of his story home. I kept it in my hand the whole walk back, like it might blow away if I let go.

They said it was lucky we found it.

I don't think it was luck. I think he wanted to be found.

Artifact: The Blood-Stained Notebook

In 1932, a small cloth-bound notebook was discovered in a forgotten trunk at an estate sale in Pennsylvania. Inside were thirty-seven names, each matched to a burial location and personal note. The final page read:

"If no one else finds them, I have."

Analysis confirmed it belonged to a Graves Registration clerk from the 314th Infantry. It is now housed at the U.S. Army Medical Museum, a silent testament to the individual efforts of war's memory keepers.

The Tools of Sacred Labor: What the Burial Crews Carried

"They gave us gloves, rope, chalk, and tags. But they never gave us what we really needed." —GRS Private, Meuse-Argonne, 1918[7]

The men assigned to Graves Registration did not carry rifles into the field. They carried something heavier: the tools of sacred labor.

By 1918, the U.S. Army had begun to standardize the equipment issued to burial details. Tucked into canvas rolls or olive-drab pouches, the contents were simple, grim, and deeply functional. There were no medals in these kits. No maps of the front. Only the objects needed to transform a body into a record, and a field into a grave.[8]

A typical burial kit included:

- Canvas body bags—crude, durable, often insufficient. When bags ran out, men improvised with tent canvas, blankets, or shelter halves.[9]

- Identification tags and wire—for attaching whatever information could be salvaged to whatever remained.

- Rope or webbing straps—for carrying remains that had stiffened in the cold or come apart in the heat.

- Chalk and grease pencils—for marking temporary wooden crosses that might stand for days or might be obliterated by the next barrage.

- Entrenching tools—the same spades soldiers used to dig trenches now dug graves. The irony was not lost on anyone.

- Rubber gloves—in theory. In practice, they were chronically scarce. Many men worked barehanded, their skin cracked and infected from contact with decomposing flesh.[10]

- Burial forms—the official paperwork that turned a body into a file: name, serial number, unit, cause of death, location of grave. Each form was a lifeline to a family waiting at home.[11]

- Waterproof notebooks and ledgers—for the information that wouldn't fit on a form or that a man wanted to remember himself.

- Lime or chloride of lime—for controlling decomposition and the spread of disease. The smell of lime became, for many GRS men, the smell of the war itself.[12]

In the field, these supplies were rarely enough. Men improvised constantly. Cigarette tins became containers for wedding rings and watches pulled from pockets.

Helmet liners carried personal effects. Bootlaces tied hands together so arms wouldn't splay during transport.[13] One burial party, finding themselves without bags or canvas, wrapped bodies in the parachute silk of a downed observation balloon—the only clean cloth they could find.[14]

The kit was functional. But for many GRS personnel, it became something more: a ritual chest.

Some men added items of their own. A small cross, a pocket New Testament, a scrap of linen to lay over a face before the dirt closed in.[15] Chaplains carried their own parallel kits containing communion wafers, consecrated wine in unbreakable vials, rosaries, prayer books in multiple languages. These were the tools for the work that came after the body was in the ground, when the words were spoken and the silence was observed.[16]

One surviving kit from the Meuse-Argonne sector, now held in the National WWI Museum in Kansas City, still bears the stains of its use. The leather strap is discolored. The notebook inside contains forty-three names, each written in the same careful hand, each followed by a grid reference and a date.[17] On the final page, in pencil so faint it nearly faded, someone wrote:

"Tell them we tried. Even when we couldn't find the pieces."

The armies of 1917 and 1918 carried weapons to make war. The men who buried the dead carried kits to end it one body at a time, with whatever reverence mud and chaos would allow.

These tools were never adequate. They were never meant to be. No kit can hold what burial demands: the capacity to see, to touch, to name, to remember.

But the men who carried them did the best they could with what they had.

And that, in the end, was all the dead could ask.

The Power of Paper: Ledgers, Names, and Records

"The war wrote its story in blood. We tried to rewrite it in ink."—Army clerk, 1919

If "baptism by mud" was the soldier's passage into death, then documentation was the soul's final attempt to return home.

By 1918, the U.S. military understood that it wasn't just waging a war of attrition or geography—it was also fighting a war of memory. With tens of thousands of Americans dying across Europe, the burden of proof—the name, the place, the cause—became essential not only for military recordkeeping, but for national mourning.

The solution was imperfect but urgent: a new layer of war bureaucracy called Graves Registration Records.

Each burial was to be recorded. Each name written, each site mapped.

The central tool for this effort was the Graves Registration Card, a paper rectangle that became a lifeline between battlefield death and home-front grief.[18] These cards listed a soldier's name, serial number, unit, date of death, condition of remains, and burial location. Some included physical descriptions of notations about personal effects.

Some were typed, others scribbled in pencil on damp field forms. Sometimes a photograph was attached. Sometimes a button or scrap of cloth was pinned to the margin.[19]

It was part ledger, part confession, part elegy.

The men who filled out these forms were not always record clerks by trade. Many were reassigned infantrymen, chaplains, junior officers, or medics. They jotted notes while kneeling beside corpses. They wrote by candlelight in crumbling barns, in trench dugouts, on the backs of ration boxes. Rain blurred names. Blood soaked corners. But still, they wrote.

And the effort was not just logistical. It was sacred.

Identification was not guaranteed. Dog tags were often missing, torn away by shellfire, or corroded beyond legibility. Makeshift burial markers—rifles jammed into the dirt, bayonets with tin plates tied to them—were vulnerable to the elements and further combat.[20] In some sectors, no markers survived. Only maps drawn in haste and hope.

Yet the Graves Registration men persisted. They developed field methods to preserve information: duplicate copies of cards, ledgers stored in waterproof oilcloth, small field maps indicating burial zones. By the war's end, entire units were devoted not just to burial, but to cataloging the war dead's final resting places.

"Everybody was a mystery you had to solve," one GRS man wrote. "And you prayed you got it right. One mistake meant a mother never got to weep at the right grave."[21]

The pressure was immense. Especially when the remains were incomplete—fragments of bone or equipment, a boot with a foot inside, a jawbone with gold fillings that had to match a dental record. These pieces were still people. And their stories still deserved to be told.

In these moments, recordkeeping became reverence.

Each accurate entry meant a telegram sent. Each marked grave meant the possibility of closure. Each card became a promise: *he is not lost.*

The emotional toll was rarely acknowledged. Few clerks spoke of the weight they carried, of writing the same cause of death over and over until it lost all meaning. Or of finding a photograph in a bloodied wallet and not knowing whether to include it or keep it, just for a moment, as a human link.

For some, this paperwork became a quiet form of coping. A way to impose order on the incomprehensible and honor the ones they could not save.

By the time the Armistice was signed, over 70,000 Graves Registration Cards had been created by American forces.[22] They would form the backbone of future military burial systems, influence the American Battle Monuments Commission, and shape the design of national cemeteries in Europe and the United States.

But more than that, they offered something fragile and necessary: *Proof that someone remembered.*

Buried Twice: The Postwar Reburial Program

"We buried them once with honor. Now we were digging them up again, with grief."—Graves Registration Service veteran, 1920

The war was over. But for the caretakers of the dead, it was only the beginning.

In 1919, as Europe counted its losses and the world tried to reimagine peace, the U.S. War Department confronted the overwhelming dilemma of what to do with the more than 100,000 American bodies buried across foreign soil.

These initial wartime graves—hurried, temporary, often unmarked—were never meant to be final. Shell craters had become burial pits. Crosses made of tin and twine had vanished in wind and rain. Entire cemeteries had sunk in flooded trenches or been flattened by the next round of artillery.[23]

Now, with the guns silenced, the government offered families a choice: leave the fallen where they lay, in a newly established permanent American military cemetery in Europe, or have them disinterred and brought home for burial in U.S. soil.

It was a monumental decision. And for many families, it was agonizing.

Most had never seen the places where their sons had died. Letters had stopped months earlier. Some didn't know if the grave even existed anymore. For many, the only connection they had was the name on a card or a folder telegram stamped "Killed in Action."

Still, tens of thousands opted for reparations. Others, especially those of recent immigrants or devout Catholics, felt it was sacred for the body to be returned to native ground.[24]

The U.S. military responded with what would become one of the largest and most emotionally charged logistical operations in American history: the Return of the Dead Program.

Between 1919 and 1922, specially trained Graves Registration teams were dispatched across Europe to locate wartime burial sites, verify identities, exhume bodies, and prepare them for transport.[25] These crews returned to the very fields where they had buried comrades just a year or two earlier, now tasked with reopening the earth they had once blessed shut.

Conditions were gruesome. Many graves had deteriorated. Flooding, animal disturbances, erosion, or battle damage had shifted markers or obliterated them entirely. Teams relied on maps, ledgers, and personal memory to find individual graves. Even then, verification was not always possible.

Bodies were exhumed with as much care as circumstances allowed. Remains were inspected for identity confirmation through dog tags, uniform fragments, personal items, dental records.[26] Each soldier was placed into a sealed metal casket, then into a wooden shipping crate marked with military precision. These boxes traveled by rail to port cities, and from there, aboard funeral ships, often met at home by solemn processions, grieving families, and flag-draped ceremonies.

But the experience was not universal.

White officers were often prioritized. Black soldiers' families, and the families of the poor, reported delays, lost paperwork, and vague or incomplete notifications.[27] Some were never told their loved ones had been buried at all.

The caretakers, the GRS men, felt the weight of this second burial more acutely than anyone. For them, the act of disinterment was not procedural but personal. Many returned to fields they remembered by name: Belleau Wood, the Argonne, Chateau-Theirry. They recognized the landmarks. They remembered the dead.

"I knew the tree. I knew the boots. It had to be him," one GRS sergeant wrote in 1921. "Digging him up again, God, I never thought we'd have to do that."[28]

Reparations brought mixed reactions from families. For some, seeing the casket arrive at a train station brought closure, a moment to grieve, to lay flowers, to mourn in a familiar churchyard. For others, it reopened trauma that had just begun to scar. The war, they said, had followed them home in a box.

In total, over 46,000 American dead were returned to U.S. soil. The rest, over 30,000, remained in newly created overseas military cemeteries across France, Belgium, and the United Kingdom. These cemeteries became hallowed ground, designed to mirror the dignity of Arlington but shaped by European soil.[29]

Yet even these resting places bore the memory of exhumation. Beneath every marble cross was a story that began twice: once at death, and once at reburial. Beneath each lay the unseen hands of men who had to reopen graves, verify names, repack remains, and offer dignity all over again—often without recognition or rest.

Their job was not done with peace. Their duty continued into the bureaucracy of mourning. And though history records the ceremonies, the returns, and the formalities, we must also remember this:

Some soldiers were buried once.

Others were buried twice.

But all were remembered by those who carried the weight.

Interlude: The Mother's Letter

"Please write and tell me if you found anything in his pocket."

The letter was creased so many times it nearly tore when I opened it. It had been forwarded from four different offices before it landed on my desk.

It read:

"My son was Corporal Thomas J. Elliott, 1st Division. Killed near Soissons.I was told he was buried there but I never received anything back.Please, if he had a photo or a letter in his pocket, can you send it to me?Even a button. Anything.I just need to know he was real at the end.I just need to hold something that touched him."

It was signed:

"Mrs. Willa Elliott, Chicago, Illinois. A mother."

I didn't find the letter until a year after we buried him again—reburied him, I mean—in Meuse-Argonne. But I still remember that boy. I remember the name stitched in red thread on the inside of his collar.

So I wrote her back.

"Yes," I said. "He had a photograph. I am enclosing it now.

And a button from his jacket.

"I held onto it, just in case someone asked."

THE WOMEN WHO CARRIED THE GRIEF

While the battlefield may have belonged to men, the mourning—quiet, relentless, and necessary—belonged to women.

World War I marked a rupture not only in the scale of global conflict but in the traditional boundaries of gender and grief. On both sides of the Atlantic, women stepped forward, not just to nurse or knit, but to tend, bury, identify, and remember. They performed the work no one else would do, and they rarely received credit for it. They were the carriers of sorrow and the custodians of legacy.

In the aftermath of battle, it was often women who washed the bodies, wrapped what was left in clean cloth, and stitched names into scraps of linen. Some were trained nurses. Others were volunteers. Many were simply nearby, village women in France, Belgium, and Italy whose fields had become cemeteries overnight.[30]

They created graves when there were none.

They marked crosses when the army hadn't arrived.

They wrote letters to mothers of soldiers they'd never met—because they had seen their boys die.

In France, rural women buried foreign soldiers with care, often praying over bodies in languages they did not speak. Some left flowers. Some left food. Some returned year after year to tend the graves. Their names are not carved in marble. But without them, entire sectors of the front would have become anonymous.[31]

In Britain, the Voluntary Aid Detachment (VAD) drew young women, many from privileged backgrounds, into work they had never imagined. Before the war, these women had been shielded from labor, from bodies, from death. Now they drove ambulances through mud and shellfire, emptied bedpans, and assisted surgeons in operating theaters thick with blood. Many VADs worked in casualty clearing stations, where the line between nursing the wounded and preparing the dead blurred into a single exhausting duty. Vera Brittain, whose memoir *Testament of Youth* would become one of the war's defining accounts, described washing bodies, folding uniforms, and writing to families—work that left her, she wrote, "old before I had ever been young." The VADs were volunteers, unpaid or barely paid, yet they performed labor as essential as any soldier's. Their proximity to death gave them a knowledge that few would share: what it looked like, what it smelled like, what it demanded of those who witnessed it.

In America, grief turned to action. The war catalyzed a surge in female-led remembrance organizations, including the Red Cross, the Gold Star Mothers, and hundreds of local memorial committees that helped compile lists of the dead, raise funds for monuments, and preserve the artifacts of war.

The American Red Cross, which employed thousands of women overseas and at home, played a key role not just in treating the wounded but in documenting the dead. Nurses and clerks kept rosters of who had died, where, and when. Some were tasked with organizing personal effects, while others ensured the correct name was attached to the right body.[32]

One of the most significant figures was Clara Barton, whose post-Civil War work in the Office of Missing Soldiers created the foundation for future recovery and documentation efforts. Though she had retired by World War I, her legacy endured in the ethos of the women who picked up the work she had begun, resolving to name the lost, to connect the pieces, to deliver peace even when hope was dim.[33]

In 1929, the Gold Star Mothers Pilgrimage gave bereaved American mothers and widows the chance to visit the graves of sons and husbands buried overseas. These pilgrimages were solemn, structured, and deeply symbolic, an official recognition that grief was a national burden rather than a private shame.[34] For many women, it was the first, and only, change to say goodbye.

Black women, however, were often excluded from this national narrative. While Black soldiers buried the dead in France, their mothers and wives were denied Gold Star Pilgrimage invitations.

Their grief, like their sons' service, was systematically erased from public commemoration.[35]

Still, they remembered. They held wakes without bodies. They embroidered names on pillows and altar cloths. They read scripture aloud and claimed their loved ones with candles and silence.

Whether at home or abroad, known or unknown, women shaped the geography of mourning. They stood at the crossroads of memory and labor, unacknowledged archivists, unrecorded chaplains, and invisible caretakers of war's emotional aftermath.

They remind us that military remembrance is not only built with uniforms and ranks; it is also built with thread and prayer, with handwritten letters and kitchen-table rituals. They bore not only the weight of the dead but the burden of ensuring that someone, somewhere, remembered.

In a war of steel and gas, it was the women who brought the softest things: cloth, memory, song.

And because of them, grief became record.

Sorrow because structure.

Love became history.

The Red Cross at the Edge of Death

"We did not carry weapons. We carried names."

By 1917, the American Red Cross had become an essential presence in the war—not merely as a relief organization, but as a bridge between the battlefield and the home front. Over 20,000 American Red Cross nurses served overseas during the war, along with thousands of volunteers, clerks, and auxiliary workers. Their duties extended far beyond bandaging wounds.

They documented the dying.

In field hospitals and casualty clearing stations, Red Cross nurses were often the last to see soldiers alive. They recorded final words. They wrote letters for men too weak to hold a pen. They noted physical descriptions, unit assignments, and personal effects, information that would later prove essential for identification and notification.

When a soldier died, the paperwork did not stop. Red Cross workers helped compile casualty lists, cross-referencing names against hospital records and burial reports. They transmitted messages to families through the Red Cross communication network, often faster and more reliably than official military channels. For thousands of families, the first word of a son's death came not from the War Department but from a Red Cross telegram.

The organization also pioneered early tracing efforts. When families wrote desperate letters asking for information about missing soldiers, Red Cross workers searched records, contacted hospitals, and sometimes walked the cemetery rows looking for names. This was the same work Clara Barton had done after the Civil War, now scaled to a global conflict.

One Red Cross nurse, stationed at Base Hospital 21 near Rouen, described her work in a letter home:

"I spent the morning with a boy from Ohio who could not stop talking about his mother's garden. By afternoon, he was gone. I wrote her that night. I told her about the garden. It was all I could give her."

The emotional labor was immense. Red Cross workers were expected to maintain composure while surrounded by death. They were trained to be efficient, compassionate, invisible, and to provide comfort without becoming part of the story. But their presence at the threshold of death made them witnesses, and many carried that witness home.

After the war, the American Red Cross would formalize many of these practices into what became Restoring Family Links, a global program that continues today. The tracing work that began in France in 1918 with one nurse writing one letter to one mother would grow into an international system for reconnecting families separated by war, disaster, and displacement.

The Red Cross did not bury the dead. But they ensured the dead were not forgotten.

Freedmen, Laborers, and the Forgotten Burials

"We dug because we were told to. But we sang because we knew they needed someone to remember."—Freedman laborer, 1865

Burial during World War I was not always voluntary. And for many of the men who shouldered the dead, there was no recognition—only silence, and the weight of what was left behind.

As American troops pushed into France, the U.S. military faced the immense task of battlefield recovery and cleanup. The chaos of trench warfare, mass casualties, and disease meant the dead could not always be collected by the soldiers who had known them. And so, into this vacuum of responsibility, a familiar labor force was summoned, one America had relied on since its earliest wars.

Black men, many recently freed or descended from those once enslaved, were called to bury the dead.

Some enlisted. Others were drafted. Others were recruited on the ground in France. They were often assigned to labor battalions, where duties included digging trenches, moving supplies, and handling the unglamourous aftermath of battle. Among the most grueling of these duties was collecting, transporting, and burying the dead.[36]

They worked without fanfare, often without proper equipment. No gloves. No masks. No boots sturdy enough to stand firm in the mud-soaked soil of Verdun or the Argonne. They lifted bodies already decomposing, sometimes fragmented, sometimes mistaken for the enemy. The men performing this labor were rarely trained. They were ordered and often coerced.

Some received pay; others did not. What few received was acknowledgement.

And yet, there is evidence that many of these men approached the work not with bitterness but with reverence. They sand spirituals while they worked, often softly, rhythmically, as they lowered bodies into trenches.

They spoke names aloud when they could, or marked them in the dust with stones and twigs when official tags were missing.[37]

They buried white soldiers who would never have shared a meal with them. They folded the coats of men who had fought for a segregated army. And they did it with care, because no one else would.

This work required a different kind of courage than military glory. One rooted in dignity rather than violence, in compassion rather than conquest.

And it was not new.

Many of these men came from a lineage of graveside laborers. Their grandfathers had dug the trenches of the Civil War. Some had helped build the national cemeteries they were now excluded from. Others had grown up tending churchyards in the segregated South, learning to care for the dead with gentleness that belied their own oppression.

After the war, some were hired as caretakers and groundsmen for the new American cemeteries in Europe. They trimmed grass, planted flowers, and straightened crosses in silence. Their names rarely made it into official records. But they became the daily keepers of memory.[38]

Still, their stories have been left largely untold.

In a war that produced thousands of pages of battlefield reports and commemorative speeches, the laborers who dug the graves were often not named, not counted, not remembered.

They were asked to carry the dead. But no one asked them how it felt.

This omission is not accidental. It reflects a truth as old as war itself: that remembrance is often shaped by power, and power does not always honor the hands that do the burying.

When we honor the Civil War and World War I dead, we must also honor those who buried them—not just the officers and generals, but the anonymous laborers in the field. The ones who dug with hands blistered by tools they weren't trained to use. The ones who sang over strangers and whispered prayers no one heard.

Because they, too, carried the war.

And they, too, remembered.

The Empire's Gravediggers: Colonial Labor on the Western Front

"They came from every corner of empire—and they buried the empire's sons."

Black American soldiers were not the only laborers of color assigned to burial work in France. Across the Western Front, the great colonial powers—Britain and France—relied on a vast labor force recruited from their empires. These men dug trenches, built roads, moved supplies, and, inevitably, buried the dead.

The Chinese Labor Corps was among the largest of these contingents. Between 1917 and 1920, nearly 140,000 Chinese men were recruited by Britain and France to serve as laborers on the Western Front. They were not soldiers—they had signed contracts promising noncombat work—but the work they performed was no less dangerous. They cleared unexploded ordnance, rebuilt shattered villages, and recovered bodies from battlefields still poisoned by gas and littered with debris.

After the Armistice, Chinese laborers were assigned the grim task of concentrating scattered graves into the new permanent cemeteries. They exhumed bodies, verified identifications, and reburied the dead in the orderly rows that visitors see today. It was sacred labor performed by men who had traveled thousands of miles from home, who spoke neither English nor French, and who were often housed in segregated camps far from the communities they served.

An estimated 2,000 Chinese laborers died during their service—from disease, accidents, and the lingering effects of gas. Many were buried in France in small cemeteries marked with Chinese characters. The Noyelles-sur-Mer Chinese Cemetery, the largest, contains 842 graves. Their headstones bear inscriptions in Chinese, selected by families who never saw where their sons were laid to rest.

Similar patterns held across the colonial world. Indian laborers served in France and the Middle East. Egyptian and South African laborers supported British operations. Senegalese and North African troops under French command both fought and performed burial duties. In almost every case, their labor was essential, their conditions harsh, and their recognition minimal.

The cemeteries they built were for others. Their own graves, when marked at all, were placed at the edges—separate in death as they had been in life.

We cannot tell the story of World War I burial without acknowledging that the hands that dug most of the graves belonged to men the empires had colonized. Their names rarely appear in memorial speeches, and their faces are absent from the photographs. But without them, there would be no Meuse-Argonne, no Tyne Cot, no Thiepval.

They carried the empire's dead.

The empire did not carry them.

Interlude: The Singing Pit

"The only music we had was our own."—Verdun, 1918

They told the men to dig fast. No ceremony. No talk. Just dig.

But the soil was heavy and slick, the bodies twisted, some barely held together. The young man on the end—Private Lewis, they called him—started humming. Just low enough to make the others look.

Then Corporal Braxton joined in. And then Samuels.

Soon, five men were singing a slow, mournful spiritual as they buried the fallen.

"Swing low, sweet chariot..."

Not for themselves. For the ones they lowered down. For the ones they were told not to speak of.

A white officer shouted for quiet. The song kept going.

Because sometimes the only thing you can give the dead is the dignity of your voice.

Racial Disparities: The Forgotten Burials

"There was no ceremony. No Salutes. No Flags. Just the hush of the forgotten, digging graves for the remembered."—Addie Hunton, 1919

Black Americans served bravely in World War I, but when it came to how they were buried, remembered, and mourned, they were not served equally.

The U.S. military was deeply segregated; this segregation extended not only to living conditions, training, and assignments, but also to death. Black soldiers were disproportionately assigned to labor battalions, and within those units, often relegated to burial duties, tasked with recovering mangled corpses, digging mass graves, and clearing the carnage of battles they had rarely been permitted to fight in.[39]

Their labor was essential. Without them, the system of military burial would have collapsed under the weight of war. Yet these men—many of them volunteers, some conscripted—were rarely credited, rarely named, and often deliberately erased from the war's commemorative record.

Even within the Graves Registration Service, racial division was rigid. White GRS personnel were given tools, protective equipment, and training. Black burial details were often left with none.

They worked longer shifts with fewer breaks and were exposed to the same horrors—without the support systems afforded to their white counterparts.[40]

Sometimes, Black soldiers were tasked with burying white troops who would have refused to fight beside them. In a segregated army, where Black men were barred from serving in combat in most divisions, they were still deemed fit to bury white bodies. This was a profound moral burden, to honor in death those who had denied them honor in life.

Addie Hunton, one of two Black women assigned to the YWCA with the American Expeditionary Forces in France, witnessed these burial duties firsthand. In her account, *Two Colored Women with the American Expeditionary Forces,* she recalled:

"They buried the white dead quietly, respectfully, with no fanfare. There were no salutes. No Flags. Just the sound of shovels and the wind moving through the woods."[41]

Even in remembrance, injustice lingered.

When the war ended and the U.S. launched the Return of the Dead program, many Black families faced delays, misinformation, or silence. Paperwork was lost. Notifications arrived months later, if at all. Some never received confirmation that their sons had died, let alone where they were buried.[42]

Meanwhile, the newly formed Gold Star Mothers Pilgrimage, intended to offer grieving mothers a federally funded trip to visit sons' graves overseas, explicitly excluded Black women from the first wave of travel.[43]

The grief of Black families was deemed less worthy of public acknowledgment.

This racial divide in burial and commemoration was not an anomaly; it was a reflection of systemic inequality that reached even into the grave. White soldiers received flags, ceremonies, and marble. Black soldiers were often buried with no marker, no name, no one allowed to witness the moment they were lowered into the earth.

Even those who survived were denied the dignity of memory. Many Black burial detail soldiers returned home with no debriefing, no support, and no way to process the psychological toll of what they had done. Some tried to speak about it. Most were not believed. Others were told that their service "didn't count" because it hadn't involved firing a gun.

But they knew better. They had buried their country's dead. And in doing so, they had borne the weight of a war that denied them honor even as it demanded their labor.

Their stories deserve more than a footnote.

They deserve recognition, restitution, and reverence.

Spotlight: Addie Hunton and Kathryn Johnson – Witnesses of the Forgotten

Addie Hunton and Kathryn Johnson were among the only Black women allowed to travel with the American Expeditionary Forces in

World War I. Their book, *Two Colored Women with the American Expeditionary Forces*, bears rare testimony to the racial disparities in both care and commemoration.

They visited camps, comforted troops, and documented abuses. But they also bore witness to the burial of Black soldiers—men denied ceremony, honor, or even names.

"They were buried without prayer," Hunton wrote. "But we remembered them aloud."

Their advocacy laid the groundwork for future demands for equality in military recognition and remembrance.

Interlude: The Unequal Telegram

"They got the notice in a box. I got nothing but a rumor."

In 1920, Clara Turner stood outside the general store in Norfolk, Virginia, gripping a torn envelope that had passed through four hands before reaching her.

It wasn't official. Just a note from someone who said they'd served with her boy."He didn't suffer," it read. "He was brave."

That was all.

A white mother across town received a polished casket, a military escort, and a full obituary in the local paper.Clara received silence. Not even a flag.

Years later, someone found her son's name on a weathered cross in a field in France. By then, Clara had passed.

No one told her they found him.

Between the Lines: Informal Truces for the Dead

"War does not always silence humanity. Sometimes, it makes space for

it."—Field chaplain, AEF, 1918

Even in the blood-soaked chaos of the Western Front, there were moments when soldiers stopped fighting—not to rest, but to bury the dead.

These moments were not ordered by generals. They weren't part of strategy or sanctioned by formal treaties. They happened between the lines, out of sight, sometimes against order, always out of necessity.

They were unofficial truces for the fallen.

The most famous of these, of course, was the Christmas truce of 1914. Across stretches of the Western Front, German and British soldiers emerged from their trenches, sang carols, exchanged small gifts, and, most importantly, buried their dead. In No Man's Land—normally a place of terror and slaughter—soldiers dug graves together, sometimes helping bury the very men they had killed days before.[44]

These moments didn't last. Commanding officers on both sides quickly reasserted control. But the precedent had been set: even in war, there is room for grief, ritual, and reverence.

By the time American troops entered the war in 1917, informal burial truces were no longer celebrated. But they still happened.

Sometimes, two sides paused fire when the stench of corpses in No Man's Land became unbearable. More often, a flag—white cloth tied to a stick—was raised, and an unspoken agreement was struck that bodies could be collected without drawing fire.[45]

American, French, German, and even colonial troops participated in these quiet negotiations. There were no documents, no medals. Just shared humanity across barbed wire.

"A German officer waved his handkerchief and called out in English," one doughboy wrote in a letter. "We met in the middle. I helped them lift their dead. One of them saluted our fallen, too. Then we went back to killing each other."[46]

These truces were never permanent. They were brief and fragile, often broken by command or shattered by suspicion. But in those short hours, the war stopped. Not because peace had been declared but because the dead demanded dignity.

In some areas, informal burial protocols developed between opposing units. A designated hour would be observed for collection. Orders would circulate through whispers and hand signals. In rare cases, soldiers left small markers or notes to indicate who had been buried where.

These were not just tactical decisions. They were spiritual decisions that affirmed the idea that even an enemy has the right to be mourned. That beneath the uniform, every soldier carried a name, a mother, a home.

These moments complicated the narratives of hate and righteousness that wars depend on. They revealed what war cannot entirely kill: compassion. They created an underground culture of grief-for-the-other, a rare and radical form of empathy on a field designed to erase such things.

These truces would shape the ethos of postwar memorials, where enemy dead were sometimes reinterred respectfully, and where reconciliation was cautiously planted in soil that had once run with blood.

They were the pause button of war, pressed not by politicians, but by the grieving, the haunted, the tired.

Because sometimes, even in war, the dead demand silence.And the living answer with a shovel instead of a rifle.

The Rise of Military Cemeteries Abroad

"Here, far from home, we lay them down not as strangers, but as sons."
—U.S. Military chaplain, Meuse-Argonne, 1923

The Great War did more than redefine how soldiers died. It redefined how and where they would be buried.

With more than 116,000 American soldiers dead by the war's end, the question of how to honor the fallen loomed large. While many families begged for the return of their sons' remains, others believed they should rest where they had fallen, among comrades in the soil of liberation. And so, for the first time in U.S. history, a system of permanent military cemeteries abroad began to take shape—carefully designed, ritually managed, and politically symbolic.

At the heart of this effort was the American Battle Monuments Commission (ABMC), established in 1923 under General John J. Pershing. The ABMC's charge was clear: to create and maintain overseas cemeteries and memorials for American war dead, primarily in France and Belgium.[47]

These spaces were not improvisational like the field graves of 1917. They were intentional landscapes of memory.

Architects and landscape designers, many of whom had served in the war themselves, worked to create cemeteries that communicated dignity, equality, and permanence. Uniform rows of white crosses and Stars of David stretched across manicured lawns. Names were etched in stone. Visitor chapels and interpretive centers were built to host pilgrims from across the sea.

Each cemetery told a story—not just of a battle, but of a nation's evolving relationship with its dead.

The Meuse-Argonne American Cemetery, near Romagne-sous-Montfaucon, became the largest American military cemetery in Europe, with over 14,000 graves.[48] Its scale spoke not only to the carnage of the war but to the scale of national commitment; these men would not be buried in anonymity, nor in haste. They would be honored in perpetuity.

Other sites, like the Suresnes American Cemetery near Paris, were selected for their accessibility, allowing grieving families a place of pilgrimage. These cemeteries became both sacred grounds and diplomatic symbols of ongoing friendship between the United States and the nations it had helped defend.

Yet these cemeteries did not merely commemorate the dead. They also commemorated a shift in the ethics of war.

For the first time, a democratic nation sought to bury its dead equally—side by side, regardless of rank. Generals and privates lay under the same white markers. This was a radical departure from older European traditions, where burial reflected class and caste. Though disparities still existed, particularly for Black soldiers, who often faced delays or omissions in burial decisions, the cemetery designs nonetheless projected an ideal of egalitarian memory.[49]

These cemeteries also became sites of ongoing diplomacy.

French, Belgian, and Italian governments granted the U.S. rights to the land in perpetuity. Monuments were often codesigned by international committees, and dedication ceremonies drew diplomats, royalty, and veterans' organizations.

In this way, the cemeteries functioned as a form of soft power, cementing America's presence abroad not through occupation but through reverent remembrance.[50]

At the same time, the ABMC imposed a carefully curated aesthetic. Markers were uniform. Decorations limited. Language solemn. The tone was one of quiet sacrifice, not political critique. Many soldiers buried there had died of disease or in labor battalions, not in dramatic charges, but the narrative of valor was preserved.

Even so, the act of naming, locating, and preserving each grave was profound.

Maps were drawn. Records meticulously kept. Flowers sent from American soil were placed on foreign graves, and pilgrimages became part of family lore. For many Gold Star Mothers, the trip to Europe to see their son's resting place became a sacred journey—one they could finally make in the late 1920s and early 1930s after early racial exclusions were slowly addressed.[51]

These cemeteries remain open to visitors today, quiet, immaculate, and powerful. They reflect the weight of war and the labor of those who stayed behind to bury the aftermath. Groundskeepers, chaplains, clerks, and volunteers shaped every row.

Each cross marks not only a life lost but a decision made: to remember.

THE WATERS OF THE ATLANTIC

"The sea does not give back what it takes."—Naval chaplain, 1918

The Great War extended beneath the waves.

German U-boats prowled the Atlantic, sinking military and civilian vessels alike. The *Lusitania*, torpedoed in 1915, carried 1,198 people to their deaths—including 128 Americans whose bodies were never recovered.[52] Troop transports, supply ships, and hospital vessels all faced the submarine threat. When ships sank, the dead sank with them.

The SS *Tuscania*, torpedoed off the coast of Ireland in February 1918, was the first ship carrying American troops to be lost to enemy action. Of the 2,013 soldiers aboard, 230 died, many trapped below decks or pulled under by the weight of their equipment. Bodies that washed ashore on the Scottish island of Islay were buried by local villagers in a cemetery that still stands, tended by descendants of the fishermen who pulled the dead from the water.

The USS *President Lincoln*, a troop transport, was sunk by U-90 in May 1918 while returning from France. Twenty-six men died. The randomness of survival haunted those who lived. One sailor later wrote: "I don't know why I grabbed the right piece of wreckage. I don't know why he grabbed the wrong one. I think about it every day."

Those who survived sinkings often watched shipmates drown, pulled under by the weight of equipment, lost in the chaos of abandoning ship, or simply disappearing beneath the waves. These witnesses carried a particular burden: the memory of faces going under, of hands reaching up, of the moment when the surface closed and the lost became irretrievable.[53]

Burial at sea, when bodies could be recovered, followed traditional naval protocol. But many of the war's naval dead remained where they fell—in the Atlantic, in the North Sea, in the waters around the British Isles. They have no headstones. Their graves are currents and depths. They are among the most invisible of war's invisible dead.

The grief of their families took a particular shape. Without a grave, mourning had no anchor. Mothers traveled to coastlines, staring out at water that held their sons but would not show them. Some threw flowers into the surf. Others kept vigil on anniversaries, watching the horizon as if the sea might finally relent.

In port towns across America, Britain, and France, memorial services were held for men whose bodies would never return. Pastors spoke of "committing to the deep," as if the ocean were a kind of consecrated ground. Perhaps it was. Perhaps any place that holds the dead becomes, by that holding, sacred.

But for the families, the absence of a grave meant the absence of a destination for grief. There was nowhere to kneel. Nowhere to leave flowers. Nowhere to say, finally, *goodbye*.

The sea remembers. But it does not speak.

Interlude: The Empty Casket

"We buried the box anyway. It was all we had."

In Duluth, Minnesota, in the autumn of 1919, the Lindgren family held a funeral for their son Erik, lost when the *Ticonderoga* was torpedoed in the final weeks of the war.

There was no body. The Navy had confirmed his death but could not recover his remains. The family received his personal effects—a pocket watch, a letter he had been writing, a photograph of a girl named Clara—but nothing else.

Erik's mother insisted on a burial anyway.

The church provided an empty casket, draped in a flag. The pastor read from scripture about those who "go down to the sea in ships." The mourners sang hymns. Erik's father placed the pocket watch inside the casket before it was closed.

"He's not in there," his younger brother said afterward.

"I know," his mother replied. "But now we have somewhere to visit."

The grave still stands in Forest Hill Cemetery. The headstone reads:

Erik Lindgren 1896–1918 Lost at Sea Remembered on Land

Artifact: The Islay Memorial Book

On the Scottish island of Islay, in the small village of Port Charlotte, a leather-bound memorial book sits in a glass case at the Museum of Islay Life. It contains the names of 266 American soldiers who died when the SS *Tuscania* and HMS *Otranto* were lost off the island's coast in 1918.

Local residents recovered bodies for weeks after the sinkings, pulling them from the water and carrying them up the cliffs. They buried the dead in two cemeteries—one at Kilchoman, one at Port Charlotte—marking each grave with a wooden cross until the American government could provide permanent headstones.

The memorial book was begun by a village schoolteacher named Margaret MacNeill, who recorded each name as it was identified, along with any personal effects found on the body. Some entries include notes from family members who traveled to Islay after the war to visit the graves.

One entry, dated 1923, reads:

Pvt. James R. Collier, Missouri. Identified by ring engraved "To James from Mother." Mother visited August 1923. Wept at grave. Left flowers and a letter. Asked us to keep tending the plot. We will.

The book remains open to visitors. The people of Islay still tend the graves, more than a century later. They consider it an obligation—not to America, but to the boys themselves.

Every November 11, a delegation walks to Kilchoman and reads the names aloud.

The sea took them. Islay keeps them.

Medical Advancements from Necessity

"The battlefield was a morgue before it was a hospital."—Medica l Corps Lieutenant, Verdun Sector, 1918

Mass death forced innovation.

World War I was a crucible of suffering that compelled medicine, mortuary science, and battlefield sanitation to evolve in real time. With millions of bodies—wounded, dying, and dead—passing through the war zone, medical and burial protocols collided. Care and cleanup became indistinguishable. Doctors and gravediggers often worked side by side. The need to preserve life bled into the need to preserve dignity in death.

This section is not about surgical miracles or battlefield triumphs. It is about how necessity transformed the way the dead were treated and how the carnage of modern war accelerated changes in preservation, documentation, and sanitation that continue to shape military deathcare today.

Embalming the Preservation in the Field

Embalming was not new in 1914; it had been widely used during the American Civil War. But the scale and pace of death in World War I pushed the practice into new territory.[54]

In many cases, bodies could not be embalmed on-site. The terrain was too chaotic, the weather too destructive, and the conditions too dangerous. However, by 1918, field preservation methods were increasingly used to delay decomposition until proper burial or repatriation. Temporary chemicals, body bags (newly standardized), and rudimentary refrigeration methods were introduced where possible.[55]

In rear-area mortuary stations, partial embalming became routine for officers and the few enlisted men approved for repatriation. These practices laid the groundwork for the more advanced embalming protocols seen in WWII and later conflicts.

The Science of Death

World War I forced medicine to confront not only how people lived, but how they died.

Military doctors, nurses, and battlefield stretcher-bearers were instructed to document causes of death—sometimes on-site, sometimes post-mortem. This wasn't only about bureaucratic order but about future learning.

Which wounds were survivable? Which procedures worked? Which gases caused lingering effects?[56] Every corpse became data, and every data point became a possible answer to how the next life might be saved.

Death records were logged with as much clinical precision as living triage notes. Autopsies were conducted not just to determine cause, but to refine battlefield medicine and identify previously unseen complications from gas warfare, shrapnel, and exposure.

The result was a blurring of the line between caregiver and caretake, between battlefield medicine and forensic examination.

Sanitation as Sacred Duty

The enormous scale of trench warfare created public health crises within military units. Decomposing bodies attracted rats and disease, polluted water sources, and weakened morale. Sanitation, once a low-priority logistical concern, became a matter of life and death.

GRS units worked in tandem with engineers and sanitation corps to ensure proper disposal of bodies, especially after mass casualty events. Latrines were moved. Burial sites were mapped to avoid water contamination. Quicklime and chemical agents were used to reduce biohazard.[57]

What emerged was a new ethic of cleanliness tied to dignity: not just for the sake of living, but out of reverence for the dead. A well-kept grave, even if temporary, was both an act of health and of honor.

Battlefield Medicine Becomes Mortuary Science

Some medical personnel, unable to save the wounded, turned their efforts to treating the dead instead. This transition was emotional as it was professional.

One nurse, discharged in 1919, recalled spending her final weeks in France cataloging and wrapping bodies for transport:

"We lost them. But I could at least make sure they didn't leave looking lost."[58]

From this work came the foundations of modern mortuary military practice, in which cause of death, condition of remains, and personal effects were standardized and documented by both medical and registration teams.

Out of blood and ruin came the structures we now associate with honor:

– Clean identification

– Gentle handling

– Verified records

– Human care, even in death

World War I didn't invent these values. But it demanded they be institutionalized.

And from the hands of medica, gravediggers, stretcher-bearers, and morticians, a new science of sacred duty emerged—one that continues to guide how the fallen are handled to this day.

The Silent War: Psychological Fallout

"No one remembers the gravedigger. Only the graves."—Former Graves Registration soldier, postwar diary

They came home. But not all of them returned.

For the men who buried the dead in World War I, survival was not the end of their war. The real battle began after the shovels were set down, after the ledgers were closed. The trauma they carried did not wear uniforms or bleed visibly, but it burrowed deep, and it stayed.

Today, we would call it moral injury, the psychic wound left when a person must perform tasks that shatter their sense of ethics, purpose, or identity.[59] For many GRS personnel, the injury wasn't from what they saw—it was from what they had to do with their own hands.

They lifted body parts. Reassembled fragments. Tied names to people who no longer had faces. They heard the whimpers of the dying and answered with silence, because their orders were to recover, not rescue.

And then they were expected to go home and forget.

Unseen Wounds

The U.S. military of 1919 had no formal structure for mental health debriefing. There were no screenings, no counseling, and no language for what these men had endured.[60] "Shell Shock" was recognized only in combat troops, and even then, often dismissed as cowardice.

But the trauma of the burial crews was different. They had not seen comrades die; they had seen what death left behind, again and again. Some wrote of dreams filled with hands reaching up from trenches. Others found themselves unable to tolerate the smell of damp earth or rotting food. Many became undertakers, cemetery caretakers, or funeral laborers, not because they were drawn to death, but because they didn't know how to leave it behind.

One former GRS soldier confessed in his journal:

"The dead don't haunt me. The living do, because I can't tell them what it cost to bring their sons home."[61]

Silence at Home

Families welcomed them with pride, parades, and platitudes. But few understood the nature of their service. GRS men weren't considered "real" veterans by some. They had not killed the enemy. They had not taken ground. And yet they carried war's most enduring burden: they were the last to touch the fallen.

The expectation to resume normal life was crushing. Some married. Some withdrew. Some drank. Others wandered. Many suffered from depression, survivor's guilt, and what we now recognize as PTSD.

There were no memorials for them. No medals, no public mourning. They disappeared into society, often as quiet men with heavy eyes who flinched when asked what they did in the war.

Coping Through Ritual

And yet, some found solace, not in forgetting, but in rituals of remembrance.

A few returned to France and took jobs maintaining the cemeteries they had once help build. Others carved crosses in their garages, kept boxes of buttons or tags, or wrote the names of the dead in personal ledgers as a way of not letting them go.

Their trauma became memory. Their memory became ritual. And ritual became survival.

The military did not recognize their wounds, but they carried them all the same. And in carrying the dead, they had sacrificed something invisible, something vital.

Their war never made the history books. Their pain never made the front page. But they were there, in the trenches, in the fields, in the ledger pages stained with rain and grief.

They were the ones who laid the foundation for how we remember our war dead.

And we owe them more than silence.

The Unnecessary: Deaths After the Armistice Was Signed

The Armistice was signed at 5:10 a.m. on November 11, 1918. It would take effect at 11:00 a.m.—the eleventh hour of the eleventh day of the eleventh month. For reasons that historians still debate, the delay was built into the agreement. Six hours of continued war, after the war was already over.

During those six hours, approximately 2,738 men died.[62]

Some commanders ordered their troops to hold positions, waiting out the clock. Others did not. American General William Wright ordered an assault on Ville-devant-Chaumont in the final ninety minutes of the war. The attack gained nothing—the position would be surrendered at 11:00 regardless. But it killed over 300 men.[63] British forces attacked at Mons, where the war had begun four years earlier, losing men in the symbolic capture of a town that would have been theirs by noon anyway.

The deadliest American action of the day was the assault on Stenay. General William Summerall wanted the town because it had bathing facilities for his troops. He ordered the attack knowing the Armistice was signed. Dozens of Americans died so that other Americans could take baths in a town that was never defended.[64]

At 10:59 a.m., men were still falling. At 11:00 a.m., the guns stopped. The silence was absolute.

And then the burial parties went to work.

The Graves Registration soldiers who gathered the bodies of November 11 knew what the ledgers would confirm: these men had died after peace was certain. Their deaths were not the fog of war, not the chaos of combat, not the terrible arithmetic of military necessity. They were the result of decisions made by commanders who chose to spend lives in the war's final breath.

One GRS clerk, processing the death certificates, noted the times: 10:30 a.m., 10:45 a.m., 10:58 a.m. He was recording men who had died after their generals knew the war was over. The weight of that knowledge settled into him. It never fully lifted.[65]

Private Henry Gunther of Baltimore is often cited as the last American killed in World War I. He died at 10:59 a.m., one minute before the Armistice, charging a German machine gun position. The Germans, who knew the war was ending, tried to wave him off. He kept coming. They fired.[66]

Gunther was buried by men who had been counting the minutes until peace. They buried him knowing he had chosen to charge. But the others—the hundreds who died because their officers ordered them forward in those final hours—were buried by men who understood that choice had been taken from them.

The physical labor was the same: recovery, identification, interment. But the moral weight was different. These bodies carried an additional burden. They were evidence of something that went beyond the ordinary tragedy of war. They were evidence of waste, of ego, of the willingness of men with authority to spend the lives of men without it even when spending accomplished nothing.

After the war, families of the November 11 dead petitioned for investigations. Why had their sons been ordered forward when the outcome was already decided? Some officers were questioned. None were punished. The deaths were recorded as combat casualties, indistinguishable in the official statistics from those who had died in the years of necessary fighting.[67]

But the Graves Registration soldiers knew the difference. They had read the times on the certificates. They had buried the last men to die in a war that was already over.

That knowledge became part of what they carried home.

Carried by the Living

"The war ended, but the work never did."

World war I shattered the old world. In its place, it left the wounded living and the unburied dead. But in the shadows of its trench lines and across the fields in France, a quiet army rose, not with rifles, but with stretchers and shovels, ledgers and cloth, silence and resolve.

They were the caretakers of the aftermath.

Graves Registration soldiers. Nurses, Laborers. Clerks. Mourners. Mothers. Gardeners. Volunteers. People whose names never appeared in victory speeches, whose war stories were left out of parades, but whose work gave the war meaning beyond destruction.

They did not fight to kill.

They fought to preserve.

To name.

To bury.

To remember.

Theirs was not a war of conquest. It was a war of compassion, fought one body at a time.

They stood between the battlefield and the home front, tasked with turning chaos into closure, turning a boy's final moments into a name that could be spoken at a kitchen table two oceans away.

Some stitched names onto scraps of linen. Some whispered prayers in languages not their own. Some dug through rain and rats to find the shape of a missing friend.

Their work was dirty. Traumatic. Holy.

And though their hands were the first to touch the fallen after battle, and the last to let go, they were often left unrecognized, unaided, and unheard.

Yet they endured.

Because remembrance is a moral act.

Because burial is not bureaucracy—it is ritual.

And because no one else would do it.

From these efforts emerged a new world of mourning.

- Cemeteries that stretched like stitched wounds across Europe
- Ledgers full of names that made grief official
- Rituals and records that shaped the century to come

They gave birth to the modern military burial system that taught the state how to grieve. They also taught us that remembrance requires labor, and that the ones who labor for the dead carry a kind of heroism that rarely wears a uniform.

And still today, in every headstone, every quiet march, every folded flag, their legacy remains:

Not just the fallen are sacred.

The ones who remember them are sacred too.

And yet in France, in Belgium, in corners of fields that no one visits anymore, there are still graves marked only: "A Soldier of the Great War—Known Unto God."

No name. No regiment. No hometown. Just a body that someone buried, a stone that someone carved, and a silence that has lasted a hundred years.

There are still men who were never found. Bodies that sank into the mud of Passchendaele, of Verdun, of the Somme, and never surfaced. Families who waited their whole lives for word that never came. Mothers who died still asking.

And there are still men who came home carrying what they could not name. Who hung a shovel in a barn and never spoke of it, who flinched at the smell of damp earth, who dreamed of hands reaching up from trenches and woke without anyone to tell.

The war ended.

The graves remained.

The caretakers scattered—into factories, farms, funeral homes, and silence.

And twenty years later, their sons would be asked to do it all again.

Intermission

Between Wars: 1918-1939

"We weren't just burying the war. We were still sorting it out."—U.S. Quartermaster Office staffer, 1931

The First World War ended in 1918. But for those who handled the dead, the war didn't stop—it shifted.

Throughout the 1920s and 1930s, quiet work continued behind the scenes. America's Graves Registration Service evolved from a reactive wartime unit into a standing bureaucracy of memory. What had begun in trenches and chaos became a structured system of global remembrance.

Clerks spent years cataloging tens of thousands of graves across Europe. Engineers redesigned cemetery landscapes for drainage, erosion control, and permanence. Artists refined headstone inscriptions. And administrators coordinated with foreign governments to maintain sacred sites—particularly in France, Belgium, and Italy, where American cemeteries had become pilgrimage destinations for grieving families.[1]

It was a time of logistical reverence.

The 1920s also marked the largest peacetime repatriation effort in

U.S. history. More than 45,000 American war dead were disinterred from their temporary European graves and returned to American soil upon family requests.[2] The rest remained overseas—by necessity or by design—watched over by local caretakers, U.S. stewards, and stone angels whose gaze never shifted.

In 1923, the American Battle Monuments Commission (ABMC) was founded to standardize and oversee this expanding system of remembrance. Under the leadership of General John J. Pershing, the ABMC symbolized a quiet but profound transformation. Memory had become national policy, not just personal grief.[3]

And yet, the system was not without its failures. Segregation still stained the process. Black families often received late or incomplete information. Burial teams remained racially divided. Commemorative literature often left Black, Indigenous, and immigrant soldiers unnamed. Grief, like service, was filtered through the lens of race, class, and silence.

Still, the machinery of remembrance kept turning.

Letters were written. Graves tended, cemeteries mapped, flowers placed. The stonework of memory was laid with careful, bureaucratic hands.

The 1930s brought no new war—but it brought no peace, either, for those who still buried the past. Veterans battled with illnesses. Families wrestled with absence. And the caretakers—the quiet force who had carried the war dead—remained at work, filing, digging, naming.

Because they understood something others had forgotten:

Peace does not erase the dead. And those who bury war do not stop just because the guns go quiet.

Symbols of Vigil: The Service Flag and Its Transformation

While clerks filed death certificates and engineers surveyed cemetery plots, another kind of caretaking was happening in American homes. In windows across the nation, a new symbol had appeared, one that transformed private grief into public witness.

In 1917, Captain Robert L. Queisser of the 5th Ohio Infantry designed the Blue Star Service Banner to honor his two sons serving on the front lines.[4] On a white field bordered in red, a blue star marked each family member in active service. On September 24, 1917, an Ohio congressman read into the Congressional Record: "The mayor of Cleveland, the Chamber of Commerce, and the Governor of Ohio have adopted this service flag. The world should know of those who give so much for liberty. The dearest thing in all the world to a father and mother—their children."[5]

The banner spread rapidly. National Geographic and Popular Mechanics ran articles about the service flag, and within months, blue stars hung in windows from Maine to California. [6] The banners were small—typically about eight by fourteen inches—but their message was visible to every passerby.[7] *Here lives a family that has given. Here waits someone who tends to hope.*

Then the telegrams began to arrive.

In May 1918, President Woodrow Wilson endorsed a request from the Women's Committee of the Council of National Defense: mothers who had lost a child in military service could cover the blue star with a smaller gold star, or replace it entirely.[8] The transformation was quiet and devastating. A gold star in the window meant that hope had ended. A son or husband would not return. The family had become caretakers of memory.

The gold star was not just a symbol of loss but a declaration of ongoing labor. Gold Star mothers did not simply grieve—they organized. In 1928, twenty-five mothers met in Washington, D.C. to establish the American Gold Star Mothers, Inc.[9] Their mission was to obtain mutual support, but it was also one of advocacy: they wanted the right to visit the graves of sons buried overseas.

The Pilgrimages: Sacred Journeys and Segregated Ships

On March 2, 1929, President Calvin Coolidge signed legislation authorizing the War Department to sponsor pilgrimages to European cemeteries for mothers and widows whose sons and husbands had been buried overseas.[10]

Congress allocated five million dollars.[11] The Office of the Quartermaster General determined that 17,389 women were eligible. Over the next four years, 6,693 would make the journey.[12]

The pilgrimages were solemn, structured, and deeply personal. Cemetery staff decorated graves with the flags of both the United States and the host country. A chair was placed beside each headstone so mothers could sit and reflect. Each pilgrim received a photograph of herself at her son's grave, where she also laid a memorial wreath.[13] For many women—aged between sixty-one and sixty-five on average—it was the first and only chance to say goodbye.[14]

But the pilgrimages carried a stain that would outlast the decade. The War Department announced that African American women would travel in separate groups, on separate ships, with separate accommodations. The law had made no mention of race. The Quartermaster Corps imposed it anyway.[15]

In spring 1930, a group of fifty-five Black mothers and widows from twenty-one states petitioned President Herbert Hoover. "When the call to arms came from our government in 1917," they wrote, "mothers, sisters and wives, regardless of race, color or creed, were asked to give their loved ones to the end that the world might be saved for democracy. This call we answered freely and willingly."[16] The War Department denied their request to travel with white women on May 29, 1930.[17]

The NAACP and the Chicago Defender urged Black mothers to boycott. Walter White, executive secretary of the NAACP, held a press conference in New York City just as the first ship carrying white women sailed out of the harbor.[18]

Approximately twenty-five women heeded the call to refuse.[19] But 168 African American mothers and widows—some elderly, some in poor health, all in grief—chose to go anyway.[20] They needed to see their sons' graves. They needed to touch the headstones. The injustice of the passage could not outweigh the necessity of arrival.

The first ship carrying Black pilgrims departed New York on July 12, 1930. Colonel Benjamin O. Davis, at the time the highest-ranking Black officer in the military, accompanied the women as an escort. Fluent in French, he volunteered to protect their dignity and ensure they received every available courtesy.[21] In Paris, famous African American expatriates staged variety shows for the pilgrims. The women dined at elegant restaurants and were honored in official ceremonies. Many returned to the United States disputing the negative coverage in the Black press.[22]

But the segregation had revealed something America preferred to forget: that the democracy for which Black soldiers had died did not extend to their mothers. That grief, like service, was filtered through the lens of race. That even in pilgrimage—the most sacred form of caretaking, the crossing of an ocean to tend a grave—Jim Crow traveled alongside them.

The Flag-Draped Casket: Final Honors

The service flag spoke from home front windows. But another flag spoke at journey's end.

The custom of draping a flag over a soldier's casket can be traced back to the Napoleonic Wars of the late 18th century. On battlefields where chaos and carnage made identification difficult, flags covered the dead so each side could recognize its own.[23] What began as practicality became ritual. By the First World War, the flag-draped casket had become the final embrace of the nation around those who had served.

The protocol was precise: the flag was placed over a closed casket with the blue field at the head and over the left shoulder of the deceased, as if the flag were embracing the body one last time.[24] During World War II, the War Department issued formal specifications for both manufacture and display.[25] After the playing of Taps, the flag was carefully removed, folded into a triangular shape, and presented to the next of kin.

The folding itself became ceremony. Thirteen crisp, precise folds reduced the flag to a tight triangle of stars—no red or white stripes visible, only the blue field. The shape echoed the tricorn hats worn by soldiers of the Revolutionary War.[26] Each fold, according to tradition, carried meaning: the first for life, the second for belief in eternal life, the third as tribute to the veteran departing. Later folds honored mothers, fathers, faith, and country.[27]

The honor guard members who performed this ritual were caretakers of the highest order. Their hands shaped not just fabric but meaning, and the flag they folded was not merely presented—it was transferred from the custody of the military to the custody of the family. The weight of service passed from uniformed hands to grieving ones. It was the last tangible thing the nation could give, proof that someone had served, someone had died, and someone still remembered.

The blue star in the window. The gold star that replaced it. The flag that covered the casket. The folded triangle pressed into trembling hands. These were the symbols of vigil, woven through the interwar years and beyond. They marked the labor of waiting, the transformation of loss, and the ongoing work of tending to what war had taken.

Because flags are not just national symbols.

They are instruments of care.

Reflection

In the last days of my grandfather's life, he finally spoke.

He was ninety-three years old. He had been a glider pilot in World War II—nineteen years old when he'd flown into combat, been captured and held as a POW, and released when the war ended. For seventy years, he had said almost nothing about what he'd experienced. The war was a locked room in his mind, and he had lost the key.

But as his body failed, something opened. The walls came down. And he began to talk. He told me about the gliders—flimsy aircraft made of wood and canvas, towed behind powered planes and released over enemy territory to crash-land with troops and supplies. He told me about being captured, about the POW camp, and about the hunger and fear and uncertainty.

And he told me about gathering the pieces of his comrades.

After crashes and firefights, someone had to collect what remained. Sometimes there were bodies. Sometimes there were only fragments—an arm here, a leg there, pieces that had been human beings minutes before. "You pick up what you can," he said. "You try to make them whole again." His voice was steady as he said it. Seventy years of practice had taught him how to speak of horror without breaking. But his eyes told me what the words could not.

"It's something you never forget," he said.

He died a few years later. But he had finally spoken. And I had finally heard. This chapter is for him and for all the nineteen-year-olds who were asked to carry what no one should have to carry—and who held it inside for a lifetime.

CHAPTER FIVE

WORLD WAR II: GUARDIANS OF THE FALLEN

A GLOBAL WAR, A GLOBAL GRAVE

They weren't surgeons or generals. Most had never seen a dead body before the war. Some had barely left their hometowns. But there, amid the barbed wire and blood of Europe, they became the keepers of grief." —U.S. Army Quartermaster Corps, 1943

World War II was not just the deadliest conflict in human history. It was the most dispersed, the most logistically complex, and the most demanding in terms of how nations confronted the aftermath of death. With over 400,000 American military dead, and tens of millions worldwide, World War II forced the United States to develop an unprecedented system for burying, identifying, honoring, and remembering the fallen.

This was not just a military challenge but an ethical reckoning, a global labor of remembrance.

From the beaches of Normandy to the jungles of the Pacific, from the deserts of North Africa to the villages of Italy, soldiers died in places no map could prepare for. And in those same places, others followed—not with guns, but with shovels, ledgers, stretchers, brushes, and prayer books.

This chapter is about those others.

The caretakers.

The namers.

The quiet ones who stayed behind to make sure the war moved forward without leaving its dead.

Following the foundations laid during World War I, the U.S. military entered WWII with a more formalized system for handling its war dead. The Quartermaster Corps' Graves Registration Service was reactivated and expanded dramatically.[1] Training became standardized, identification protocols were sharpened, and maps were drawn before battles were fought. Burial teams accompanied infantry units, not as afterthoughts, but as embedded specialists prepared for what came next.

But the system was never perfect.

Battlefield conditions often overwhelmed the best-laid plans. Pacific humidity reduced remains to bone in days. Tank battles left entire crews incinerated. Bombed ships disappeared beneath waves. And in Europe, front lines shifted so quickly that graves were sometimes abandoned before they could be marked.

Still, the caretakers kept going.

They worked through gunfire. They carried corpses under shell bursts. They pieced together fragments of identity from shattered dog tags, wedding rings, and burned Bible pages. They did this not just in Europe, but in jungles, deserts, mountains, and islands where the war never stopped.

And as the war raged, the task of remembrance grew. Temporary cemeteries sprang up on every front, and engineers laid out row upon row of makeshift graves with meticulous geometry. Chaplains held services in foxholes. Nurses helped write final letters. Artists sketched the dead so their mothers might see a peaceful face.

And as in previous wars, not all who labored for the dead were honored equally.

Black soldiers, Indigenous troops, women, and Jewish and Japanese-American GIs all served in burial duties and memory work, often in segregated conditions. Their grief was heavy; their recognition minimal.

By the war's end, the United States would face a second monumental task: the Return of the Dead program. It would become the largest patriation effort in military history, moving more than 170,000 remains from across the globe to permanent resting places either at home or in newly constructed American military cemeteries overseas.[2]

This chapter traces that journey from battlefield to burial. From chaos to order. From forgotten to remembered.

It is not only a story of systems.

It is a story of people, those who carried the grief, the stretchers, the ledgers, the memory.

They are the guardians of the fallen.

And this is the war they fought.

Interlude: The Letter They Never Sent

"I came here to fight. They gave me a pen, a shovel, and silence."

Private Ellis Morgan, age 20, landed on a Normandy beach with his unit three days after D-Day. He never fired his weapon. Instead, he was reassigned to Graves Registration.

Tucked inside his chest pocket, he kept a notebook filled with letters he never sent. Letters to mothers, to wives, to children. Letters to the dead.

One read:

"I don't know your name. But I carried you out. I found a watch on your wrist and made sure they didn't lose it. I tied your boots together so they wouldn't fall off when we moved you. I said a prayer I hadn't remembered since I was ten."

Ellis never sent the letters.But he remembered every name he wrote down.

And when the war was over, he brought the notebook home. Someone had to carry the stories.

Drafted to Bury: The Men of the GRS

"I didn't even know what Graves Registration was. Thought I was going to cook."—Private First Class, 603rd Quartermaster Graves Registration Company

They didn't enlist to bury the dead.

Many of the men assigned to the GRS during World War II had never heard of such a unit until the orders came down. Some had signed up as musicians, cooks, clerks, or engineers. Others were draftees handed a shovel and a ledger instead of a rifle, told simply: *you'll understand when you get there.*

The GRS existed within the U.S. Army Quartermaster Corps, charged with the recovery, identification, and burial of American war dead. But while the unit had its origins in bureaucratic aftermath of World War I, it was World War II that transformed it from a paperwork appendage into a frontline force of sacred logistics.[3]

By 1943, entire companies (such as the 603rd, 604th, and 612th Quartermaster Graves Registration Companies) were deployed into every major theater: Europe, North Africa, the Pacific, and the China-Burma-India front.[4] These units were dispatched not just in response to mass casualties, but in anticipation of them.

Yet for all their importance, they were rarely glorified.

Combat troops often viewed GRS men as unlucky, unclean, or unnerving. Their presence was a grim reminder of what might follow. Some soldiers avoided them entirely, while others whispered about the things they saw. GRS men lived in liminality, not quite among the living, but not yet gone.

And it began, always, with the moment they were chosen.

Drafted Into Death

In some cases, selection for Graves Registration was random. A man showed aptitude for organization or stoicism, and a superior officer marked him for reassignment. Other times, it was the absence of visible trauma—a soldier who had not broken under stress—that marked him as suitable to face the dead.

There was rarely a choice.There was rarely preparation.

One GRS veteran remembered, "They asked if I could write neatly and stomach blood. I said yes. That was my mistake."[5]

Once assigned, most received only minimal training. They were taught how to:

– Tag bodies with identification markers

– Fill out casualty reports

– Handle remains respectfully

– Lay out temporary cemeteries

– Record GPS-like field coordinates by hand

– Practice sanitation to prevent the spread of disease

But nothing in the training prepared them for the smell of burned flesh, or the sound of a sobbing friend recognizing a corpse by the ring still stuck on a swollen finger.

What they learned, they learned while digging. In motion. Under shellfire. Surrounded by flies and silence and grief.

Psychological Selection, Unofficial Survival

Some men adapted quickly, developing emotional detachment as a defense. Others collapsed within weeks. Others, still, developed strange rituals to survive: writing down the names of the dead in private ledgers, keeping photos or trinkets until they could let go, or whispering prayers over the nameless.

Behind the humor and hard eyes, many broke.

"I stopped seeing faces. I only saw jobs. I had to. Or I wouldn't be able to keep digging," wrote one GRS corporal in a letter never mailed.[6]

They were expected to be efficient, clinical, professional. But no amount of training could prepare a man to stitch together the pieces of a fellow soldier. Some GRS personnel were issued psychological evaluations, especially by 1944, but few received real mental health support.[7]

Some drank. Some prayed. Some wrote poems in secret. Some never spoke of the war again.

War Without Pause: Burial on the Battlefield

"We lived in death's shadow—and we didn't stop working just because the sun went down."—GRS technician, European Theater, 1944

There was no front line that spared the dead.

No road far enough to escape them.

From the hedgerows of Normandy to the peaks of Monte Cassino, from Pacific atolls to the frozen forests of the Ardennes, the Graves Registration Service followed every American advance—not with weapons, but with tools of memory: stretchers, shovels, tags, typewriters.

Their mission was relentless: recover the bodies, identify the remains, bury the fallen. Repeat.

Geography as Enemy

Burial on the battlefield was shaped by terrain as much as tactics.

In the Pacific, humidity meant corpses decomposed within twenty-four hours. The stench arrived before the bodies could be reached. Jungle undergrowth, disease, and monsoons made every step a risk. In Italy, bodies tumbled down steep hillsides riddled with mines. In France, mud clogged every boot print, and hedgerows swallowed the dead whole. In Belgium, during the Battle of the Bulge, frozen ground turned shovels into splinters.

"The ground was full of the dead, but there was no time to stop and grieve," one GRS soldier wrote. "We carried them out like broken branches, because there were too many to count."

No Pause for the Dead

Combat didn't stop for burial. GRS units often worked under fire—during artillery lulls, between air raids, or under sniper watch. Their movements were strategic, timed with commanders, but they were rarely guaranteed safety. In many cases, they were forced to bury the dead in provisional mass graves, using bulldozers, tarpaulins, and crude markers.

Some bodies were buried multiple times as the front shifted. Some were never recovered at all.

Temporary cemeteries sprang up with whitewashed markers, wooden crosses, and handwritten field ledgers. In theory, every grave was recorded with coordinates and serial numbers. In practice, chaos ruled. Tags went missing. Serial numbers were miswritten. Paperwork was soaked or burned.

Still, they tried.

Still, they believed in the ledger.

Rituals in Ruin

Chaplains could not always be present. On those days, GRS men improvised.

– They sang softly, or not at all.

– They rested a hand on the coffin before lowering it.

– They whispered "you're home now" to a boy they would never know.

They learned to bury without ceremony—but not without care.

Some graves were dug in churchyards ruined by shellfire. Others beside tank roads, or in clearings where only days before, trees stood tall. In forward areas, GRS soldiers stacked caskets against the walls of their tents due to lack of space. They slept beside the dead, listening to the drip of rain and waiting for the sun.

They carried portable typewriters. They mapped grave plots in oilskin notebooks. They set up folding desks in barns or chapels with stained-glass windows shattered by bombs. Every recovery followed the same routine:

1. Recover
2. Identify
3. Inventory
4. Bury
5. Report

Some days, they buried dozens.

Other days, hundreds.

And on the worst days, thousands.

There were no medals for this work. No parades.

But there was devotion—not to nation or victory, but to the idea that no soldier should vanish without witness.

"Burial on the battlefield," one GRS officer wrote, "was a kind of prayer. Not spoken. Not sung. But dug into the ground with reverence."

And so, they kept digging.

A Sacred, Stigmatized Duty

Despite the trauma—and perhaps because of it—many GRS soldiers developed a deep reverence for their duty.

They weren't warriors in the traditional sense. But they were something else: guardians of the fallen. Stewards of final memory.

"We didn't shoot bullets. We stitched people back together long enough for their mothers to bury them," said a veteran of the 612th GRS Company.[8]

They were the last to see the fallen. The last to touch them. The last to speak their names out loud.

They bore witness so that others could grieve.They recorded the dead so they would not vanish.

And though few volunteered for this task, many came to believe it was the most important thing they had ever done.

They were drafted to bury.But they chose to remember.

The Pacific Theater – Bodies and Distance

"In the jungle, the dead didn't wait. They disappeared."

If burial in Europe was a race against chaos, burial in the Pacific was a race against nature itself.

The Pacific war scattered American dead across thousands of miles of ocean, on islands whose names most Americans had never heard: Guadalcanal, Tarawa, Saipan, Peleliu, Iwo Jima, Okinawa.[9] The distances were staggering—thousands of miles from the nearest major port, weeks from home by any measure. And the conditions made every aspect of burial exponentially harder.

Heat and humidity accelerated decomposition. In tropical climates, a body could become unrecognizable within twenty-four to forty-eight hours.[10] Dog tags corroded. Paper records dissolved in monsoon rains. Photographs curled and faded. Graves Registration personnel learned to work fast—faster than in Europe—or lose the ability to identify remains entirely.

The jungle consumed evidence. Bodies left overnight were often found partially buried by vegetation or scattered by animals. In some cases, remains were never recovered at all—swallowed by swamp, washed out to sea, or lost in terrain so dense that search parties could pass within feet and never see them.[11]

Island warfare produced its own horrors. On Tarawa, the assault on a heavily fortified Japanese position killed nearly 1,000 Marines in seventy-six hours.[12] Bodies lay on the beach for days, baking in the sun, while the battle raged. When GRS teams finally reached them, many could only be identified by location: the third man in the third row of the charge. Serial numbers were matched to positions, and positions were matched to after-action reports. It was forensic archaeology performed in real time.

The sea claimed its own dead. Naval battles—Midway, Leyte Gulf, and countless smaller engagements—sent thousands of sailors to the ocean floor. These men received no burial in any traditional sense.[13] Their ships became their tombs. Their names were recorded, but their bodies were never recovered. For their families, closure meant accepting that the Pacific itself was the grave.

Temporary cemeteries on Pacific islands were hastily constructed and constantly threatened. Some were shelled after being established. Others were overrun when positions changed. At Guadalcanal, one cemetery was relocated three times before the island was secured.[14] Engineers learned to dig deeper, mark more carefully, keep duplicate records in separate locations—lessons paid for in lost identities.

The emotional distance compounded the physical distance. For families on the American mainland, the Pacific was incomprehensibly far. Letters took weeks. Telegrams were terse. The landscape—palm trees, coral beaches, volcanic rock—bore no resemblance to anything familiar. When bodies did return, often years later, families received remains from a world they could not imagine.

And many bodies never returned at all.

The Pacific Theater would produce more long-term missing-in-action cases than any other front.[15] Decades later, the Defense POW/MIA Accounting Agency (DPAA) continues to recover remains from Pacific islands—emerging from jungle growth, uncovered by erosion, discovered by construction crews or local villagers. The work that GRS teams began in 1942 is still not finished.

For the men who performed burial duty in the Pacific, the conditions were uniquely brutal. They worked in heat that caused heatstroke. They handled remains in advanced decomposition without adequate protective equipment. They fought disease—malaria, dysentery, tropical infections—while performing labor that weakened their bodies and haunted their minds.[16]

One GRS sergeant, stationed on Saipan, wrote in a letter he never sent:

"The smell doesn't leave you. It's in your clothes. Your skin. You can shower until your hands bleed and still smell it at night. But you keep working. Because if you don't, they disappear. And someone's mother will never know where her son is."

The Pacific war tested the limits of what burial systems could accomplish. It proved that identification was not just a bureaucratic function but a moral imperative—and that even the best systems could not recover everyone.

The islands kept their dead, and the families waited.

Some are waiting still.

The Ships That Became Tombs: Burial at Sea in World War II

In the Pacific Theater, the ocean claimed the dead on a scale never seen before.

Pearl Harbor set the pattern. On December 7, 1941, the USS *Arizona* exploded and sank in less than nine minutes, killing 1,177 sailors and Marines. The ship's hull became their tomb, a mass grave that remains undisturbed to this day.[17] The Navy determined that recovery of the remains was impossible without destroying the memorial that had formed around them. And so, the dead stayed where they fell, entombed in steel and silt, still technically on active duty because they have never been recovered.

The *Arizona* was not unique. Across the Pacific, ships went down with their crews—torpedoed, bombed, or destroyed in the chaos of naval warfare. The USS *Indianapolis*, torpedoed after delivering components of the atomic bomb, sank in twelve minutes in July 1945. Of the 1,196 sailors aboard, approximately 900 survived the sinking—only to face sharks, dehydration, and exposure in the open ocean for four days before rescue. Only 316 survived.[18] The bodies of the dead drifted and disappeared, taken by the sea and its predators.

For those killed in action aboard ship, burial at sea was immediate and necessary. Naval vessels carried body bags and canvas, and chaplains performed services daily during periods of heavy combat. The dead were prepared, prayed over, and committed to the deep, sometimes while the ship was still under attack.[19]

"We buried men between battles," one sailor recalled. "The guns would go quiet for an hour, and we'd gather on deck and say the words and let them go. Then the guns would start again, and we'd go back to our stations."[20]

The ritual provided structure, but it could not provide closure. Families at home would learn their sons had been "buried at sea," a phrase that conveyed both honor and absence. There was no grave to visit, no stone to touch. Only the Pacific itself, vast and unknowable, holding the dead in its depths.

Today, the USS *Arizona* Memorial spans the sunken battleship, allowing visitors to look down into the water where the dead still rest. Oil continues to leak from the wreck—the "tears of the *Arizona*," some call it—a reminder that the ship and her crew are still present, still unrecovered, still awaiting a homecoming that will never come.[21]

The memorial receives over a million visitors each year. Many of them are descendants of the men entombed below. Some throw leis into the water. Others stand in silence, searching for something they cannot name.

The Pacific war produced hundreds of such underwater graves. Ships are scattered across thousands of miles of ocean, each one a cemetery that cannot be tended. The dead of the sea war have no headstones. Their memorial is the water itself.

For the sailors who survived—who watched shipmates go down, who performed burial rites between bombardments, who lived knowing they might be next—the Pacific carried a particular weight. The ocean was not just the battlefield; it was the grave. And every wave was a reminder of what lay beneath.

The Ledger and the Face: Identification and Recordkeeping

"We weren't just clerks. We were the last link between the battlefield and someone's kitchen table in Kansas."—GRS corporal, 1945

Every soldier carried a name. It was the job of the Graves Registration Service to make sure that name was not lost.

If burial was the act of laying a body to rest, identification was the act of returning the soul home. The process looked bureaucratic—filled with paper, pens, fingerprints, photographs—but it was sacred. For every unmarked corpse on a field in Belgium or Saipan, there was a family in Texas, Pennsylvania, or Wyoming waiting for answers. For every mistake, there was a telegram delivered to the wrong door.

In World War I, identification had been erratic and improvised. World War II changed that. With standardized records, advances in forensic practices, and a greatly expanded Quartermaster Corps, the U.S. military developed the most comprehensive identification system in its history to date.[22]

When a body could be reached, GRS personnel conducted an immediate on-site search:

– Dog tags were collected (one left with the body, one for records)

– Pockets were emptied, personal effects bagged

– Teeth, scars, and tattoos were noted

– Unit insignia, equipment, and clothing were documented

– Items were carefully packaged and cross-referenced

But in the fog of war, tags were often missing. Fire, water, or explosion made identification elusive.

So they learned to look harder.

Dental records became the primary standard.[23] Each soldier had charts on file. In cases where facial recognition was impossible, jaws and teeth became the language of truth. Fingerprints were sometimes used, though unreliable in field conditions. And when both failed, the smallest details—a locket, a handwritten note, a sewn laundry mark—became the only clues.

Each case was logged on a Graves Registration Form, listing:

– Physical characteristics

– Cause of death (if known)

– Grave location

– Serial number

– Personal effects recovered

When possible, a photograph of the deceased was taken prior to burial. These were filed with the War Department and cross-verified through local command.

Temporary cemeteries were meticulously mapped, with grid coordinates and reference numbers designed to prevent dislocation if the front moved.[24] Still, chaos intruded. Cemeteries were bombed, and grave markers rotted. Sometimes, entire sections of remains had to be recovered and relocated multiple times.

"Behind every cross is a story," one GRS captain wrote, "and behind every story is a stack of papers we hoped never had to be written again."[25]

Artifact: The Dog Tag – A Small Metal Promise

"It wasn't the metal that mattered. It was what it carried."

The military identification tag—colloquially known as the "dog tag"—is often seen as a utilitarian object: stamped metal, name, rank, serial number, religion. But in war, it became something more. It was a tether between life and death. A token of identity when all else was unrecognizable.

By World War II, every American service member was issued two tags. One remained with the body, and the other was collected for recordkeeping.[26] But soldiers quickly gave them another purpose.

They called them "buddy tags."

Worn in pairs, they were not only military issue but personal talismans. Many men wrote blood types or next-of-kin on the back. Others etched initials or added tiny photos. Some wore them under the shirt, close to the heart. Others looped them around their boots, just in case the torso was gone before the name could be read.[27]

In battle, identification tags became sacred.

GRS teams often found them buried in mud, clutched in fists, or dangling from shattered helmets. One recovery report from Okinawa noted a tag "still warm from the hand that held it." In Normandy, a GRS corporal described crawling through a crater to retrieve a tag that had "no body left to claim it."[28]

The dog tag was also a signal:

- To the chaplain, where to place a prayer
- To the gravedigger, what name to write
- To the family, that their son was not lost in anonymity

But they weren't always legible. Fire blackened them. Shrapnel tore them. In the Pacific, corrosion set in within days. GRS technicians learned to clean, polish, and even restamp damaged tags to recover names.[29]

After the war, many families received these tags with the returned effects. Some framed them. Others placed them beside photographs or wore them around their own necks in quiet homage.

One dog tag, now held at the National Museum of American History, belonged to a soldier killed at Anzio. The metal is dented and discolored. The stamped letters are nearly worn smooth. But someone—a mother, a wife, a child—polished it until it shone again.

A small metal promise: *you mattered. You will be remembered.*

Interlude: The Second Burial

"He dug graves with his hands, no gloves, no mask. And when he died, they forgot his name."

Private Luther Bell was part of a segregated GRS labor unit operating in the Pacific in 1944.

He wasn't given proper burial gear. His boots were too small. His mask torn. But he did his job, pulling scorched bodies from jungles and digging graves in the rain day after day.

He wrote one letter home before dysentery took him:

"Ma, I don't think anyone knows what we do out here. But it's work that matters. I try to make it holy. Even if no one sees it."

His unit buried him in a temporary cemetery with no marker. Later, they couldn't find him.

Years passed. His name didn't make the official list of honored dead. His family received no flag. No ceremony. Only silence.

Some say he was buried once in the Pacific.Then again beneath history.

Spotlight: The Nisei Mortuary Workers – Burying the Nation That Imprisoned Their Families

"They served a country that didn't trust them. They buried the dead of an army that questioned their loyalty."

Among the least known stories of World War II burial work is that of the Nisei—second-generation Japanese Americans—who served in the U.S. Army's mortuary units while their families remained imprisoned in American internment camps.

Beginning in 1942, over 120,000 Japanese Americans were forcibly removed from their homes on the West Coast and incarcerated in camps scattered across the interior of the country.[30] Yet even as the government declared them security risks, it continued to accept Japanese American volunteers and draftees into the military. More than 33,000 Nisei served during the war, many in the celebrated 442nd Regimental Combat Team, the most decorated unit in American military history.[31]

But some Nisei served in less visible roles, including the Quartermaster Corps units responsible for handling the dead.

These men—many of them trained in mortuary science before the war, or assigned because they could "stomach the work"—performed the same duties as other GRS personnel: recovery, identification, preparation, burial. But they did so carrying a particular weight. Their parents and siblings were behind barbed wire in Manzanar, Heart Mountain, or Tule Lake. Their citizenship was questioned. Their loyalty was doubted.

And still, they buried the American dead.

One Nisei mortuary worker, interviewed decades later, described the contradiction:

"We were handling the bodies of men who died for a country that put our families in camps. I didn't know what to feel. Angry? Proud? Mostly, I just felt tired. And I knew those boys deserved to be treated right, same as anyone. So I did my job."[32]

The irony deepened when Nisei workers handled the remains of soldiers from units that had included men who might have supported—or even enforced—the internment orders. They buried without distinction. They treated every body with the same care. They did not allow the injustice done to their families to compromise the dignity they owed the dead.

After the war, many Nisei veterans returned to communities that had been scattered by internment. They rebuilt lives amid prejudice that would take decades to fully dissipate. Most never spoke publicly about their mortuary work. It was too painful, too complicated, too deeply entwined with the contradictions of their wartime experience.

In 1988, the U.S. government formally apologized for internment and provided reparations to surviving internees.[33] But no specific recognition has ever been extended to the Nisei who served in mortuary units—men who performed sacred labor for a nation that had betrayed them, who never stopped believing that the dead deserved better than the living had received.

They remain among the war's most invisible caretakers.

They buried the contradiction.

And they carried it home.

THE HUMAN COST OF ACCURACY

Despite the overwhelming scale—over 400,000 U.S. military

deaths—the Graves Registration Service achieved a stunning identification rate. By the war's end, fewer than 5% of American casualties remained unidentified.[34]

That accuracy came at a price.

Some men spent hours—or days—on a single body. Comparing teeth to files, matching ring inscriptions to home addresses, praying that the watch they found was truly his. Mistakes did happen. But the GRS implemented triple-check systems: unit confirmation, medical cross-checks, and itemized records to prevent devastating errors.

They were clerks, yes—but also detectives. And when they weren't sure, they said so. A wrong name was worse than none at all.

They faced ethical dilemmas, too. What if the effects suggested one name, but the dental chart didn't match? What if two bodies were found together, their tags missing, and one wallet between them?

There were no protocols for those days. Only judgment and silence.

"We decided by what felt most right," a sergeant recalled. "And then we wrote it down and carried the guilt with us."

The Ledger as Memorial

The GRS was a machine of memory.But it was run by men.

Behind every typed form at the Normandy cemetery, every row of white marble in Manila, is a trail of reverent paperwork—men hunched over desks in foxholes, chapels, and tents doing the slow, painful work of holding on to names.

They weren't just preserving history.They were preserving someone's son.

War can take a life in a moment.But forgetting takes only silence

.The ledger existed to stop that.

Buried Unequally: Race, Gender, and Erasure

"The rows of white crosses stand in perfect symmetry. But symmetry is not the same as equality."—Memorial curator, Normandy, 1997

Behind the solemn order of American military cemeteries—rows aligned, names engraved, flags raised—lies a story less often told. It is a story of disparity in death, and of the quiet erasure that shaped not only who was remembered, but who was allowed to do the remembering.

For all its commitment to honor and procedure, the Graves Registration Service was not immune to the structural inequalities of its time. Race, gender, and class shaped the wartime dead—not just in how they were buried, but in how they were valued.

Black Soldiers, Invisible Labor

Black Americans served in every branch of the military during World War II. Over one million Black men and women wore the uniform, and many were assigned to Quartermaster units, including those responsible for burial.[35] These segregated labor battalions were the backbone of GRS operations, tasked with recovering bodies from the frontlines, digging mass graves, moving caskets, and maintaining temporary cemeteries.

Yet their contributions were systematically underrecognized, underequipped, and under supported.

Black GRS personnel were often given:

- Inferior tools and inadequate protective gear
- The most hazardous burial assignments
- Limited access to chaplains or spiritual services
- No authority, even over white soldiers with less experience

"They buried the bodies that white America wanted to forget—both the fallen and themselves," wrote a post-war GRS observer.[36]

In some cases, Black soldiers buried white soldiers who would not have fought beside them in life. Even in death, segregation held. And when the war ended, these same men were excluded from memorial services, denied veteran honors, or left out of the official record altogether.

Many of these laborers died in service. Some, like Private Luther Bell, were buried without markers, their names lost in field reports or misfiled documents. Their families often received no word, or notification too late to grieve with dignity.

Atrocity and the Limits of Burial

"There were no graves. There was only ash."

This book is about the caretakers of war dead—those who recovered, identified, buried, and mourned. But World War II confronted the world with something that defied all categories of care: industrialized annihilation designed to leave no bodies, no names, no graves.

The Holocaust was not a battlefield. It was an erasure.

Six million Jews, along with millions of Roma, disabled persons, political prisoners, and others deemed undesirable by the Nazi regime, were murdered in a systematic campaign of extermination.[37] Many were shot in mass executions and buried in pits across Eastern Europe—at Babi Yar, Ponary, Rumbula, and hundreds of other sites.[38] Others were transported to camps like Auschwitz-Birkenau, Treblinka, Sobibor, and Belzec, where death was administered with factory efficiency.

At these sites, the dead were not buried. They were burned.

The crematoria operated continuously. Bodies were reduced to ash and scattered—in rivers, in fields, in ponds—specifically so that no trace would remain.[39] The Nazis understood what this book has argued throughout: that burial is identity, that a marked grave is a claim on memory. By destroying bodies entirely, they sought to destroy memory itself.

There were no dog tags. No ledgers. No crosses or stones.

Only absence.

And yet, even in this void, caretakers emerged.

Survivors became the first witnesses. Those who lived through the camps—emaciated, traumatized, barely alive—returned to mass graves and cremation sites to mourn what could not be recovered.[40] In the weeks after liberation, displaced persons gathered at pits and ruins to say Kaddish, the Jewish prayer for the dead, over ground that held no identifiable remains. Some survivors dug. At Treblinka, where the Nazis had attempted to destroy all evidence, Jewish survivors and Polish villagers returned after the war to sift through earth mixed with bone fragments and ash.[41]

They could not identify individuals. They could only acknowledge that people had died there and that the ground was sacred because of what it held, even if it could never be named.

Allied forces, too, became reluctant caretakers. When American and British troops liberated camps like Dachau, Buchenwald, and Bergen-Belsen, they encountered death at a scale that overwhelmed all training.[42] GRS units were not prepared. There were no forms for mass graves. No protocols for bodies stacked like cordwood. No procedures for identifying the unidentifiable.

At Bergen-Belsen, British forces used bulldozers to bury the dead in mass graves—not out of disrespect, but because disease was spreading and there was no other way.[43] Army chaplains spoke prayers over pits while soldiers wept. Photographers documented what words could not convey.

General Dwight D. Eisenhower ordered that German civilians from nearby towns be marched through the camps to witness what had been done.[44] He understood that the world would someday doubt. He wanted witnesses.

But witness is not burial. And burial, in the traditional sense, was impossible.

What do caretakers do when there are no bodies to care for?

They remember anyway.

In the decades after the war, Holocaust memorialization took forms that traditional military cemeteries could not. Yad Vashem in Jerusalem became a repository of millions of names gathered through testimony and research, each one a recovery of identity that the Nazis had tried to erase.[45] Memorial sites at Auschwitz, Treblinka, and other camps preserved the grounds themselves as graves—because the ground was all that remained.

These were not burials in any conventional sense. They were acts of sacred witness, acknowledgments that human beings had died, even when their names could never be known, even when their bodies had been turned to smoke.

This section sits uneasily in a book about military burial. The Holocaust was not a battlefield, and its victims were not soldiers. But it belongs here because it reveals the outer limit of what burial can do—and what happens when that limit is deliberately, systematically destroyed.

The caretakers of World War II did not only bury soldiers.

They also confronted the abyss.

They stood at the edge of pits and crematoria and understood that some wounds cannot be closed, some dead cannot be named, some grief has no grave to visit.

And still, they bore witness.

Because even when burial fails, remembrance must not.

The Liberator's Burden: When Care Became Impossible

The soldiers who liberated Nazi concentration camps were not prepared for what they found. No one could have been.

When American forces entered Buchenwald on April 11, 1945, they encountered piles of emaciated corpses stacked like cordwood. Survivors, barely alive, reached toward them from bunks packed five high. The smell—that particular smell of mass death—hit them before they crossed the gates and would never fully leave their memories.[46]

When British forces entered Bergen-Belsen four days later, the scale was even more staggering: an estimated 10,000 unburied bodies and 60,000 survivors in various stages of starvation and disease. Typhus was rampant. The camp was so contaminated that traditional burial was not merely difficult—it was medically impossible.[47]

The decision was made to bulldoze bodies into mass graves.

This was not cruelty but necessity. The alternative—leaving bodies to decompose in the open while proper graves were dug—would have accelerated the typhus epidemic and killed thousands more. Military doctors made the calculation: mass burial now, or mass death from disease. They chose the living over the dignity of the dead.

But the soldiers who drove the bulldozers were not doctors making calculations. They were nineteen-year-olds from Leeds and Liverpool, from Kansas and Kentucky. They climbed into the cabs of military vehicles and pushed human bodies—hundreds of them, then thousands—into trenches. They watched limbs bend at wrong angles. They saw faces disappear beneath other faces. They did what they were ordered to do because someone had to do it.

And they carried that weight for the rest of their lives.

"I still see them," one British soldier said, fifty years later. "I see them in the blade of the bulldozer. I see them falling into the pit. I know we had to do it. I know there was no other way. But God help me, I pushed them in."[48]

The SS guards who had created this horror were made to participate in the burial. At Bergen-Belsen, surviving guards were forced to carry bodies by hand into the mass graves—a form of rough justice, perhaps, but also a practical necessity given the scale of death. Some guards collapsed from typhus. Some were beaten by survivors. Some simply walked into the graves and lay down.[49]

But even forcing perpetrators to share the labor could not lift the burden from the liberators. They had inherited a landscape of death created by years of bureaucratic murder, industrial killing, and deliberate starvation. They had arrived too late to save most of the dead and too early to process them with dignity. They occupied the impossible middle ground as witnesses to atrocity, conscripted into its aftermath.

General Dwight D. Eisenhower, upon touring Ohrdruf concentration camp, ordered that as many American soldiers as possible be brought to see the evidence.

> "We are told that the American soldier does not know what he is fighting for," Eisenhower said. "Now, at least, he will know what he is fighting against."[50]

He also ordered that German civilians from nearby towns be marched through the camps to let them see. Let them bear witness. Let them understand what had been done in their name, near their homes, while they claimed not to know.

The witnessing was necessary. It created the evidentiary record that would be used at Nuremberg. It forced confrontation with atrocity that might otherwise be denied. But for the soldiers doing the witnessing—and the burying—it was also traumatic. They were asked to absorb what no human being should have to absorb.

Many never spoke of it afterward. The silence of the concentration camp liberators became its own phenomenon—a collective muteness that lasted decades.[51] When they finally did speak, often in old age, the details came flooding out: the smell, the piles, the bulldozers, the faces.

The faces most of all.

These soldiers were caretakers thrust into an impossible role. They did not create the dead. They did not run the camps. They arrived as liberators, as saviors, and found that salvation had come too late for tens of thousands. What remained was disposal. Processing. The grim logistics of mass death.

They did the work. They bore the witness. And they carried it home to countries that celebrated victory without fully understanding what victory had required them to see.

The Holocaust's caretakers were not the SS guards who ran the crematoria. They were the young men who came after, who cleaned up what hatred had made, who pushed bodies into graves because there was no other way.

They deserve to be remembered. Not as perpetrators. Not as heroes. But as witnesses conscripted into testimony, carrying evidence in their memories that the world would need when justice finally came.

Interlude: The Typing Pool

"We typed faster when the names were young. I don't know why. Maybe we wanted them home quicker."

They sat in rows, fingers flying over Remingtons, their backs straight, their heels together. In a converted office in Washington, dozens of women typed names for hours.

NAME: PVT. THOMAS EDWARD BANKSCAUSE: KIA, NORTH AFRICANEXT OF KIN: MRS. LUCY BANKSPERSONAL EFFECTS: DIARY, RING, COMB, BIBLESTATUS: DECEASED

A supervisor once asked one of them to slow down. Her typing was too forceful, keys sticking.

She didn't stop.

"I wasn't angry," she later said. "Just trying to keep up with death."

The Invisible Women: Clerks, Caregivers, and the Burden of Bureaucracy

"They told us we were just typing names. But we were grieving with our fingers."—U.S. Army Graves Registration clerk, 1945

The iconic images of America's war dead returning home—flag-draped coffins, solemn escorts, marble headstones—often overlook a crucial part of that journey.

Before the ceremonies, before the graves, before the folded flags, there were ledgers. And the women who filled them.

During World War II, as millions of men were deployed to the front, women stepped into roles across the home front. While many are remembered as factory workers or nurses, others served quietly in the clerical wings of the Graves Registration Service (GRS), processing the scale of death through paper, pen, and persistence.[52]

Their mission was emotionally surgical:

– File over 280,000 official death records– Match remains to dental records and fingerprints– Cross-reference personal effects with serial numbers– Mail out medals, watches, and last letters

"We processed death like it was inventory," one clerk remembered. "But every number was a boy. Sometimes it felt like we were drowning in ghosts."

Women Behind the War's Final Curtain

Some clerks were enlisted members of the Women's Army Corps (WAC), trained for burial support. Others were civilian hires under federal contract.[53] They worked long hours in administrative centers across Europe, North Africa, and the United States, many stationed in postwar depots coordinating repatriation and final disposition.

Their tasks were repetitious, exacting, and emotionally grueling:

- Proofreading next-of-kin letters
- Verifying matches between grave markers and field records
- Sorting through sealed trunks of personal effects
- Filing last photographs into "Final Correspondence" folders

In effect, they became the final link between the fallen and the families who waited.

Care Through Paperwork

These women rarely saw the bodies. But they handled everything else:

- A wristwatch with dried blood
- A Bible with a mother's note inside
- A wedding band with initials carved into the inside of the gold

They weren't asked to cry. They were asked to type.And type.And type.

"I couldn't wear perfume for years," one WAC later said. "The smell reminded me of the trunks. Shaving kits. Cologne. The scent of boys who didn't come back."[54]

They mailed medals. Folded flags. Sealed letters. They read the letters that came back from widows and mothers asking, *was he scared? Was he alone? Can you tell me what happened to his ring?*

They were told not to answer personally.But some did anyway.

Gendered Silence and the Women Who Served

Of the 9,386 Americans buried at the Normandy American Cemetery, only four are women. All were nurses.[55]

This number reflects not only the demographics of combat, but the systemic exclusion of women from formal military honors in death. Thousands of women served in uniform—WACs, WAVES, Army Nurses, Red Cross volunteers. Some died in air raids, shipwrecks, or while tending to the wounded on the front. Yet most were not formally recorded by the GRS system.

Their burials were often left to happenstance or local authorities. Some were interred with military precision. Others were omitted from rosters or buried under civilian status despite having died in uniform.

Their stories rarely appear in official reports.But their losses were real.

Unequal Access to Remembrance

Even among men, rank and class affected burial outcomes.

- Officers were more likely to be identified, repatriated, and buried with full honors
- Enlisted men—especially from rural areas, immigrant families, or with less documentation—were more likely to remain unidentified
- Families with political connections or public platforms were sometimes able to expedite returns or secure special cere-

monies

While the GRS strived for fairness, the reality of bureaucracy often reflected social hierarchies. Some bodies received folded flags, official escorts, and eulogies. Others returned quietly, anonymously, or not at all.

The disparity extended to enemy dead as well. Axis soldiers were usually buried with identification and respect per the Geneva Conventions, but often in separate, minimally maintained cemeteries, especially in the Pacific. In contrast, Allied dead (British, Canadian, French) were often interred side by side with Americans, with shared ceremonies and records.

Carriers of Memory, Denied Their Own

Despite the inequalities, those most marginalized often became the most vital caretakers of remembrance.

- Black soldiers maintained cemeteries long after the battles were over
- Female clerks and nurses recorded names by candlelight in damaged aid stations
- Immigrant laborers filed reports, carried coffins, and drafted the maps future generations would walk upon

They buried the heroes.They filed the paperwork.They carried the grief.

But their own stories were left untold.

Some were buried twice: once beneath the soil, and again beneath history.

Unrecognized and Unnamed

No plaques honor their service. No monuments bear their names. They were not photographed beside generals or interviewed after the war.

But the records they kept—the precision they preserved—became the only proof of life for thousands of grieving families.

They carried grief by keystroke.Their mourning was administrat ive.But it was still mourning.

Because remembrance is not only carved in stone.Sometimes, it is typed in silence by women who turned the last breath of a generation into ledgers of love.

The Red Cross at War's End: Bridging the State and the Family

"We were the voice between the telegram and the kitchen table."

By 1945, the American Red Cross had become an essential intermediary between the military machine of death and the families who waited at home. What had begun in World War I as improvised letter-writing and hospital visitation had matured into a comprehensive system of support, notification, and connection.

Over 100,000 American Red Cross volunteers and staff served during World War II, both overseas and on the home front.[56] Their roles ranged from distributing coffee and doughnuts to troops, to sitting beside dying soldiers in field hospitals, to the most delicate work of all: helping families navigate the bureaucracy of loss.

The Red Cross's Service to the Armed Forces division operated at every stage of the casualty process.[57] When a soldier was wounded, Red Cross workers helped transmit messages to families, often faster than official military channels. When a soldier died, they assisted with the paperwork that connected a name in a ledger to a mother's grief. They did not replace the military's notification system, but they humanized it.

In military hospitals across Europe and the Pacific, Red Cross workers performed duties that blurred the line between nursing and mourning. They wrote letters for soldiers too weak to hold a pen. They recorded last words. They sat with the dying when no chaplain was available. And when death came, they ensured that personal effects—a watch, a ring, a photograph—were properly tagged and forwarded, so that families would receive something to hold.

One Red Cross volunteer, stationed at an evacuation hospital in France, described her work in a letter home:

> *"I held his hand while he talked about his daughter. He'd never met her—she was born after he shipped out. He asked me to tell her he loved her. He died that night. I wrote the letter the next morning. I didn't know what else to do."*[58]

After the war, the Red Cross played a crucial role in the Return of the Dead program. As families made the agonizing choice between repatriation and overseas burial, Red Cross caseworkers helped them understand their options, navigate the paperwork, and prepare for what would come.[59] They served as a buffer between the impersonal language of government forms and the raw reality of grief.

For families receiving remains, Red Cross volunteers often met the funeral trains at stations. They offered support during ceremonies. They connected families with counseling resources—limited as they were in the 1940s—and with other Gold Star families who understood their loss.

The Red Cross also continued its tracing work. Families seeking information about missing service members could submit inquiries through the Red Cross network, which coordinated with military authorities and international partners.[60] In many cases, they provided answers when the War Department could not.

This work was not glamorous, and it did not make headlines. But it represented something essential: the recognition that bureaucracy alone could not serve the grieving, that families needed more than a telegram and a form letter, that somewhere between the state and the kitchen table, someone had to stand and translate loss into something bearable.

The Red Cross women and men who performed this work were caretakers, too. They did not bury the dead, but they ensured the dead were not forgotten. They did not carry bodies, but they carried messages, effects, and the weight of witnessing.

Their names rarely appear in histories of the war.

But they appear in the letters families kept—the ones that said, *he was not alone. Someone was with him. He is remembered.*

Interlude: The Train Pulled in at Dusk

"They didn't speak when the train pulled in. They just took off their hats."

In 1948, a funeral train arrived in Wichita, Kansas. Onboard was one soldier. One casket. One name.

The station had no band, no speeches. Just a family, waiting beneath the long shadow of an American flag.

A GRS sergeant in dress uniform stepped down. He handed the father a folded flag and a sealed envelope. The father nodded. The mother wept silently.

The name on the casket matched the one in the letter.So they believed it.Because they needed to.

Later, the GRS sergeant wrote in his log: *"Delivery complete. Family received remains with dignity."*

But in the margin, in smaller handwriting, he added: *"She kissed the casket. He couldn't speak."*

Coming Home: The Return of the Dead: Repatriation and Sacred Bureaucracy

"Because bringing the dead home was not just a policy—it was a prom ise."—U.S. Army Quartermaster Corps, 1948

The war was over, but the dead were still on the move.

Between 1947 and 1951, the United States undertook one of the most ambitious, complex, and emotionally charged operations in its history: the Return of the Dead Program. Over 170,000 American service members who had died in World War II were disinterred, identified, prepared, and transported back to U.S. soil—or laid to rest in permanent military cemeteries overseas, depending on their families' wishes.[61]

It was not simply a logistical mission. It was a national mourning ritual, built on ledgers and silence, policy and prayer.

A Choice Between Two Burials

Families of the fallen received a letter from the War Department asking whether they wished their loved one's remains to be returned to the United States or interred in one of the new overseas military cemeteries being built and maintained by the American Battle Monuments Commission.

The choices were not easy:

- Normandy, Lorraine, or the Ardennes in Europe
- Manila, Saipan, or Guam in the Pacific
- Or... a grave in a hometown plot, beside parents or siblings

Most families chose repatriation.[62] Others—especially those whose sons had died in decisive or liberating battles—felt peace in letting them rest where they had fallen.

The Mechanics of Reverence

Each return began with exhumation. Graves Registration personnel disinterred bodies from temporary wartime cemeteries, some buried years earlier under fire or in haste. Remains were cleaned, verified, and repackaged in uniform caskets, metal liners inside wood, sealed with care. Personal effects were cross-referenced. Identification was reconfirmed using triple-check methods.[63]

The journey was deliberate.

- Bodies were moved from disinterment centers to coastal ports
- Then loaded onto U.S. military transport ships

- Then placed aboard funeral trains, often accompanied by military escorts

Across the country, stations grew quiet when the trains arrived. Coffins draped in flags were lowered onto platforms, received by families—some in private, others in full public ceremony.

THE HOME FRONT RITES OF PASSAGE

Some funerals were private: a single grave, a folded flag, a quiet prayer.

Others were public events: marching bands, rifle salutes, newspaper headlines like *"Local Hero Returns Home."*

Regardless of form, the return became an act of civic grief—not just personal. Americans watched bodies come home and were reminded of what war had cost. Mourning became visible again, and with it, a sense of national closure.[64]

But not every return brought peace.

Some families received caskets marked "Unknown." Others questioned whether the remains inside truly belonged to their loved ones. A few were later notified that errors had occurred—a misfiled tag, a swapped report. For many Black families, the return was slower, less formal, or marred by bureaucratic silence.[65]

Inequity did not end at war's conclusion.

THE GRS CARRIES THE WEIGHT AGAIN

For Graves Registration veterans, the Return of the Dead was a second war.

Some recognized names they had recorded years earlier. Others opened caskets they had once sealed. Many returned to the same fields in Italy, France, or the Philippines where they had dug the first grave.

"We buried him once with urgency. Now we were burying him again with reverence," one GRS officer wrote.[66]

The emotional toll was immense.

Repatriation required reliving the war, one body at a time. But for those who performed this work, it was more than duty. It was devotion to a code, to a promise, to an idea that even in bureaucracy, dignity must prevail.

The Meaning of the Final Place

The new American cemeteries overseas became permanent testaments to sacrifice. Beautifully landscaped, globally negotiated, and visited by thousands each year, they remind us that memory is not always national—it is transnational, rooted in both geography and love.

And the thousands who returned home, buried in family plots, rural cemeteries, or national memorial parks, carried another message:

That America had not forgotten.

That in a war that fractured the world, something sacred had endured: the idea that no soldier dies alone.

That if a body falls, someone will come.To lift it.To name it.To carry it home.

Memory at Scale

"They came home. That's what matters."—GRS Sergeant, Repatriation Unit, 1950

World War II was not just a war of armies and empires; it was a war of names, of remains, of remembrance at scale.

The Graves Registration Service, shaped by necessity in World

War I, had, by 1945, become a mature institution. It was meticulous. Comprehensive. Global. It retrieved bodies from bombed villages, sunken ships, Pacific beaches, European forests, and frozen hillsides. It processed death with military precision and mourned with quiet repetition.

For the first time in American history, mourning became part of national infrastructure.

- Cemeteries were designed with architectural symmetry
- Ledgers were typed with decimal-like detail
- The repatriation program ran with the discipline of a postal service and the solemnity of a chapel

Memory had been scaled.

But remembrance was still incomplete.

A Burden Carried in Silence

The Graves Registration Service succeeded—logistically. But the cost of that success was incalculable.

These men—and women—buried war while it still roared. They held the fragments of youth and futures and promises. They logged them, bagged them, buried them. They mapped the graves, wrote the letters, and returned to their bunks lined with crates.

They were never granted the distance of time.They could not *move on*. Because their job was to stay behind.

They became caretakers of both the dead and the nation's collective grief.[67] And yet, they were rarely honored, rarely named, and rarely seen.

They received no medals for stitching limbs for identification. No statues for hand-copying casualty reports in bombed-out churches. No parades for stacking coffins under jungle trees or freezing tents.

Their trauma was absorbed—not discussed.Their service, buried beneath narratives of glory.

What We Owe the Rememberers

We remember the paratrooper. The nurse. The sailor. The flag-raiser.

But what of the man who placed the dog tag in a pouch?The woman who typed a boy's name into a ledger, then went home to an empty desk?The soldier who closed the lid on a casket in silence?

These people didn't die in war. But something in them did.

What they gave us was more than order. More than burial.

They gave us:

- The certainty of identity
- The location of a name
- The dignity of saying goodbye

In doing so, they made grief possible for millions.

And yet.

In 1951, a casket arrived at a train station in rural Georgia. Inside were the remains of a soldier who had died at Anzio seven years earlier. His mother, now elderly, had waited the entire time. She touched the flag and whispered, *"I knew you'd come home."*[68]

She buried him beside his father, in a cemetery that held four generations of their family. The telegram had said "remains verified." She chose to believe it.

Somewhere in the National Personnel Records Center, a file remains open. The name is known. The death is confirmed. But the body was never recovered. A family in Ohio still writes letters, still asks questions, still hopes that someday the phone will ring.[69]

Somewhere in the Philippines, a construction crew will unearth bones next year, or the year after, or the year after that. The DPAA will be notified. The slow work of identification will begin again. And a family that stopped waiting decades ago will receive a call they no longer expected.[70]

The war ended in 1945.

The burials did not.

The last transport ships carried remains home into the 1950s. The last temporary cemeteries were closed. The last GRS units were deactivated, their men scattered back into civilian life, carrying what they had seen into factories, farms, offices, and silence.

Some of them never spoke of it.

Some of them never stopped.

One former GRS sergeant, interviewed in 1985, was asked what he remembered most about the war. He paused for a long time. Then he said:

"The weight. I remember the weight. Not just the bodies. Everything. The names. The letters. The families who would never know. You carry that. You carry it until you can't anymore. And then you keep carrying it."[71]

The Graves Registration Service succeeded logistically, but the cost of that success was incalculable.

They gave families the ability to mourn. They gave the nation the ritual to grieve. They turned chaos into care.

And they were never thanked.

Not properly. Not enough.

This chapter has traced what they did. The next will follow where the work continued—in Korea, in Vietnam, in every conflict since. The systems they built would be tested again and again. The burden they carried would be passed to new hands.

But the weight remains.

It always does.

Interlude: The Night Watchman

"No one told me to light the lanterns. But I knew the dark was too heavy alone."

He wasn't a soldier. Not officially.He was a caretaker—hired after the war to tend a temporary cemetery just outside Reims.No one remembered his full name. The paperwork listed him simply as "L. Jones." Every night, he walked the rows with a kerosene lamp, checking markers, clearing leaves, humming spirituals beneath his breath.

Some nights, he paused before a cross and whispered: "Still here, son."Other nights, he lit an extra flame when the fog rolled in thick.

A visiting officer once asked why the lanterns were necessary. "No one's watching," the man said.

L. Jones looked out across the field of white wood and said quietly, "That's where you're wrong."

When the cemetery was formalized, the lanterns were removed.B ut the groundskeepers kept finding the lamps—lit—well into the 19 50s.No one confessed.Some say it was the wind catching flame.Others say it was memory refusing to go dark.

Reflection

I was traveling home with my daughter. She was barely seven years old.

We had a layover, and like so many travelers, we passed the time watching planes arrive and depart, watching luggage carts move back and forth along the tarmac.

Then one plane drew my attention.

From it, a flag-draped casket was carefully removed.

The concourse fell still. Conversations stopped. People rose from their seats and moved quietly to the windows. Together, strangers without instruction or announcement, we placed our hands over our hearts and watched as a mother and father received their child, their body returned home from a war somewhere in the world.

The casket was accompanied by soldiers who walked beside their fallen comrade with a tall, practiced reverence, each step deliberate, ceremonial, sacred.

My daughter clutched my leg, uncharacteristically silent. She did not ask what was happening. She simply stood with us, holding space in the terminal, understanding something without words.

And then, without warning, tears slipped down my face.

Chapter Six

Korea and Vietnam: New Landscapes, Same Grief

New Wars, Old Wounds

"We brought them home in silence, to a country that didn't want to hear."—Vietnam mortuary specialist.

The wars in Korea and Vietnam marked a shift, not just in how America fought, but in how it remembered.

These were wars without declarations. Wars that blurred battle lines and political purpose. Wars shaped not by surrender ceremonies or public triumph, but by stalemates, withdrawals, and contested legacies.

And for the men tasked with handling the dead, these conflicts brough not only new terrain—frozen mountains, jungles thick with rot—but new kinds of invisibility.

There were no caissons rolling through capitals, no universal parades for the fallen. There was no sweeping national unity.

Yet, still, the caretakers of death remained.

In Korea, they dragged frozen bodies from hills no one would remember. In Vietnam, they pulled fragments from mud, tagged them, and zipped them into plastic.

They flew escort beside flag-draped caskets only to return to a country that didn't want to see them.

This chapter is for them.

The ones who retrieved the fallen. The ones who were told not to speak. The ones who carried war home in their bodies, their memories, and their silence.

They didn't choose these wars.

But they buried them.

Between To Wars: Forgotten but Familiar

"I buried men in Europe. Then I buried them in Korea. Different war, same cold ground."—Former Graves Registration Service NCO, 1952 oral history

The Korean War has often been called "The Forgotten War." And for those who worked in the GRS, that title stung—not because they craved recognition, but because forgetting was the one thing they had worked tirelessly to prevent.

Only five years after the end of World War II, the U.S. military found itself once again shipping caskets home from foreign soil. But this time, the context had changed. What had once been a globally defined war against tyranny was now a limited regional conflict framed by Cold War containment. The mission was different, the stakes ambiguous. But for GRS personnel, the job was the same: recover the dead, identify the remains, document everything, and bury with care.

Once again, they were asked to do so under nearly impossible conditions.

A Harder Landscape

The Korean Peninsula presented a new and brutal geography. Unlike the more logistically accessible theaters of WWII, Korea's mountainous terrain, extreme weather, and unstable front lines made recovery and burial exponentially harder. Units advanced and retreated across the same ground within days or weeks. GRS teams operated on icy ridgelines, in mined valleys, and in territories still under sniper fire.

During the desperate early months of the war—particularly at the Pusan Perimeter and the Inchon landings—field burials were hasty, often marked with improvised wooden crosses or not marked at all. Rapid retreats meant that many of these graves were lost or destroyed. In some cases, entire units were overrun before remains could be recovered.[1]

The winter of 1950–51 was especially harsh. Subzero temperatures turned the ground into concrete. GRS units used blasting powder to soften the earth or delayed burials until thaw. Some remains were stored in refrigerated tents, caves, or makeshift morgues. One soldier recalled, "The bodies froze in the positions they died in. We buried statues."[2]

Even with improved postwar field equipment and training, many GRS personnel in Korea were veterans recalled from WWII or young draftees with minimal experience. Fatigue was compounded by repetition. It was the same labor, the same grief—but with fewer reinforcements, fewer resources, and no glory.

The Missing and the Forgotten

Korea produced one of the largest numbers of unrecovered American war dead in modern history. Over 7,500 U.S. service members remain unaccounted for. Many are believed to lie in unmarked battlefield graves, collapsed trenches, or within North Korean territory, where U.S. recovery teams have never been allowed to operate.[3]

In 1954, the United Nations Command and North Korea participated in Operation Glory, which exchanged remains across the demilitarized zone. Over 4,000 sets of remains were transferred. But many were returned without identification, and some arrived so badly decomposed or disarticulated that proper burial required GRS teams to rebuild, reinventory, and relabel every item.[4]

Each unknown was still given the full process:

– Autopsy if possible– Dental charting– Fingerprinting– Reburial with headstone– Ledger entries mailed to the War Department

But many files ended with a haunting word: "Unknown."

No National Mourning

Unlike World War II, the Korean War never inspired a national narrative of noble sacrifice. No new cemeteries were built in its name. No films canonized its dead. The conflict ended not with a declaration, but with a ceasefire—and its burial work faded from view.

GRS personnel returned to a country more interested in prosperity than remembrance. There were no victory medals for the burial corps, no monuments to the ones who mapped the graves. Their ledgers sat in filing cabinets, their maps folded in archives, their service largely forgotten.

But they had done the work.

They had dug in frozen hills.They had pulled bodies from m inefields.They had marked graves that no one ever visited.They had remembered when no one else would.

And in the shadows of World War II's triumph, Korea offered a bitter reminder:

It is easy to build systems of memory.It is easier still to ignore the memories we do not want to carry.

Interlude: Winter Ledger

"He was frozen stiff when they brought him in. I couldn't bend his arm to take the tag off. So I wrote around it."

It was January 1951. A corporal with the Graves Registration Service sat in a tent lit by a kerosene lamp. Outside, the snow drifted against the canvas. Inside, he bent over a body bag, slowly thawing on the ground.

The man inside had been found on a ridge above Chipyong-ni, frozen in a crouch, rifle still slung across his back. His boots were stuck to his feet. His jaw was locked. His name was still legible on the tag, but his face had gone blue-white.

The corporal logged the time. Sketched the outline of a body too rigid to move. His pencil snapped in the cold.

Later that night, he drank instant coffee and stared at the empty page beside the ledger.

"One more boy gone," he muttered. "One more no one will ask about."

He folded the report.

The snow fell harder.

Cu Chi and the Jungle Dead: Vietnam's Haunting Terrain

"We weren't just looking for bodies. We were looking for pieces. And sometimes we couldn't tell if they were ours."—U.S. Army Mortuary Affairs Specialist, Vietnam, 1970.

If the cold preserved the dead in Korea, Vietnam devoured them.

The jungles near Cu Chi—dense with bamboo, riddled with tunnel systems, and soaked in monsoon mud—offered no favors to the Graves Registration personnel who followed the line of fire. In a war without front lines, where the enemy could disappear underground or melt into the tree line, death came suddenly, intimately, and often incompletely.

Graves Registration teams operated in an environment where nothing was stable. A body might be booby-trapped, a patrol ambushed mid-recovery or remains half-consumed by the terrain before a team could even reach them. There were no caissons here, no hymn-lined processions through towns waving flags. There were plastic sheets in the jungle, stretchers slung in silence, and a pervasive stench that seeped into everything.

Recovery in the Unrecovered World

The region of Cu Chi, notorious for its sprawling tunnel systems, became one of the most haunting areas of operation. Recovery teams pulled bodies from collapsed shafts and tree-covered kill zones, often retrieving only fragments. "Sometimes we carried parts," one mortuary specialist recalled. "We stopped asking what we were holding."[5]

High humidity, scavengers, and tropical decay accelerated decomposition to such a degree that even basic identification became a race against time. Traditional body bags frequently failed. Many teams resorted to wrapping remains in multiple layers of plastic sheeting, carried alongside shovels, rubber gloves, and rudimentary tag kits.[6]

It wasn't just the terrain—it was the uncertainty. A soldier might disappear during a firefight, his position unconfirmed. Days later, a patrol might locate a shallow grave or blood trail that led nowhere. Recovery became detective work, filled with doubt and dread.

"We didn't call them soldiers anymore," one GRS veteran later said. "We called them cargo. It was the only way to sleep at night."[7]

The Night Shift at Tan Son Nhut

The dead left Vietnam through Tan Son Nhut Air Base.

Located on the outskirts of Saigon, Tan Son Nhut served as the primary aerial port of embarkation for American casualties throughout the war. Here, mortuary operations ran around the clock, processing, preparing, and loading remains onto transport aircraft bound for the United States.[8]

The night shift was preferred by many workers. The temperatures were cooler, the air traffic quieter. And there was something about the darkness that made the work feel less exposed, as if the dead deserved privacy in their departure.

One mortuary specialist described the routine:

> *"We worked from midnight to eight. You'd process the remains, check the paperwork three times, seal the transfer case. Then you'd watch them load it onto the plane. Some-*

times you'd stand there until the wheels lifted off. Just to make sure they were really going home."[9]

The facility operated until the final American withdrawal in 1975. In its last weeks, as Saigon fell, personnel worked frantically to evacuate remaining remains and records. Some files were lost. Some remains were left behind—to be recovered, if at all, only decades later.[10]

For those who worked the night shift at Tan Son Nhut, the memory of the fluorescent lights, the hum of refrigeration units, the weight of transfer cases, and the silence of the tarmac at three in the morning as another plane lifted into the darkness never faded.

They were the last to see the dead before they crossed the Pacific. They were the final hands in Vietnam.

The Poisoned Ground: Agent Orange and Burial's Long Shadow

The jungle was not the only enemy consuming remains. The chemicals America sprayed to destroy it would complicate burial (and death) for decades to come.

Between 1961 and 1971, the U.S. military sprayed approximately 20 million gallons of herbicides over Vietnam, Laos, and Cambodia as part of Operation Ranch Hand.[11] The most notorious of these was Agent Orange, a defoliant containing dioxin—one of the most toxic compounds ever synthesized. The goal was to deny the enemy cover and food. The cost would be measured in generations.

For mortuary workers, Agent Orange created immediate complications. Remains recovered from sprayed zones were often contaminated.

Handlers reported skin irritations, respiratory problems, and symptoms that would only later be understood as chemical exposur e.[12] There were no protocols for handling contaminated remains, and no warnings were given. Men processed bodies from defoliated areas with bare hands and standard equipment, unaware of what they were absorbing.

The long-term effects proved devastating. Veterans who had served in mortuary units—like combat troops who had walked through sprayed jungle—began developing cancers, neurological disorders, and other illnesses at alarming rates.[13] Many died young. Their deaths were not counted as war casualties, but they were war deaths nonetheless—delayed by years or decades, but rooted in the same conflict.

Agent Orange also complicated the identification of remains recovered long after the war. Dioxin persists in soil for decades, and recovery teams working in Vietnam in the 1980s and 1990s encountered contaminated sites that required specialized handling.[14] The poison that had been sprayed to reveal the enemy had also poisoned the ground where American dead lay waiting to be found.

For families, Agent Orange added another layer of uncertainty. Veterans who had survived the war only to die of exposure-related illness often faced bureaucratic battles for recognition. Their deaths were not always counted as service-connected. Their widows fought for benefits, and their children, some born with birth defects linked to dioxin exposure, carried the war into yet another generation.[15]

The dead of Vietnam were not only those who fell in combat. They were also those who came home, lived for years, and then died of what the jungle had left inside them.

And the men who buried the dead?

Some of them became casualties too—years later, in hospital beds far from the war, still carrying the poison they had unknowingly absorbed while performing sacred labor.

Death Under Fire

Unlike in World War I and II, there were no formal truces in Vietnam for body recovery. GRS teams moved under active fire across uncertain terrain. Jungle paths became ambush zones. Rice paddies concealed mines. Any noise could signal a sniper. In one after-action report, a four-man GRS team recovered five bodies under machine gun fire, unable to retrieve the sixth.[16]

They traveled light—dog tag kits, waterproof ledgers, tape, stretchers—but carried emotional weight that never lifted. In the rare lulls, they wrote names with shaking hands in ledgers wet with rain or blood.

Double Standards in Death

Enemy bodies, when recovered, were often buried hastily, burned, or left in the jungle. Viet Cong and North Vietnamese soldiers were not granted the same rituals, records, or respect. This discrepancy, often justified through military necessity, left a moral residue. Some GRS men spoke of their growing discomfort with the dehumanization. "At first, I thought it was policy. Later, I realized it was easier not to look at them as people."[17]

The war blurred every moral line.

Civilian casualties were common. Villagers were killed by misfired artillery, elders were caught in crossfire, children were buried under huts destroyed by air raids. While not under official GRS jurisdiction, these deaths were witnessed—and often mourned—by those tasked with handling American losses. Some buried civilians when they could. Others helped Vietnamese families recover bodies. But most were left only with memories. And guilt.

Erased from the Screen

Vietnam was a televised war—but not for GRS

Journalists filmed firefights. Protesters photographed returning coffins. But few ever captured the moment when a Graves Registration soldier unzipped a body bag or tried to identify a comrade by dental chart. Their work was too graphic, too real, too raw for public consumption. It didn't fit the narrative—neither of valor nor of villainy.

They existed in the margins.

And when they came home, they stayed there.

The jungle swallowed the bodies. The nation swallowed the story. And the men who carried the war's remains? They carried the silence.

Invisible Labor: Race, Draft, and Burial Duty

"Some of us fought. Some of us dug. Either way, we bled."—African American veteran, Quartermaster Corps, Vietnam War

In Vietnam, who died was political. But who buried the dead—that was even more so.

The Vietnam War unfolded at the intersection of military conflict and domestic unrest. As protests surged across the United States, so too did a national reckoning over race, class, and systemic inequality. Nowhere was that more viscerally felt than in the quiet, underacknowledged ranks of the GRS.

In Vietnam, burial was dangerous, relentless, and emotionally consuming labor. More often than not, that labor fell to the marginalized:

Black, Latino, and working-class white draftees who were statistically less likely to secure deferments or desirable military assignmen ts.[18] These men didn't carry rifles into battle.

They carried body bags.

Vietnamese Hands, American Dead

The labor of processing American remains did not fall solely to American personnel. At mortuary facilities across South Vietnam, local civilian workers performed the essential tasks of washing bodies, preparing remains, maintaining equipment, and handling the unglamorous logistics that kept the system functioning.[19]

These workers—many of them women—labored in facilities designed to process death at industrial scale. They were paid wages far below what American personnel received, working alongside chemicals and decomposing remains without adequate protective equipment. And when Saigon fell in 1975, many faced retribution for having collaborated with American forces.[20]

Their names do not appear in American military records. Their contributions are not honored in American memorials. But they, too, were caretakers of American dead, performing sacred labor for a nation not their own, in a war that would eventually consume their country.

We do not know how many there were. We do not know their names.

But they were there. And they did the work.

Systemic Inequality, Rewritten in Death

Though the U.S. military had been officially desegregated since 1948, the on-the-ground reality in Vietnam often reflected deep structural disparities. Black and brown service members were disproportionately assigned to Quartermaster roles, including GRS and mortuary operations—jobs labeled "non-preferred" for their physical demands and emotional toll.[21]

In the GRS, these disparities meant:

- Fewer promotions
- Less access to psychological care
- More frequent placement in hazardous zones
- Constant exposure to trauma with no decompression protocol

Some Black GRS veterans described being sent into contaminated recovery sites filled with decomposing remains, chemical residue, and active danger without adequate equipment or backup.[22]

And still, they did the work.

They zipped the bags. They matched the tags. They folded uniforms and cleaned medals no one would thank them for delivering.

Drafted to Dig

While middle- and upper-class white men were often able to obtain college deferments, National Guard placements, or commissions, young men from rural towns and urban centers were swept into support roles that required more muscle than prestige.[23]

Vietnam's labor divisions weren't just about race—they were also about class.

A farm boy from Alabama might find himself hauling remains up a hill in An Loc. A young Puerto Rican man might spend months in a field morgue near Pleiku, cleaning stretchers and filing death reports.

In the government's eyes, this was logistics. But for those who performed the labor, it was grief in motion, a job in which emotional numbness became the only way to survive.

"We honored the dead," one Black veteran said, "even when our own lives were dishonored."[24]

No Space to Grieve

The psychological impact of this work has rarely been studied. Many GRS veterans of color reported symptoms now recognized as PTSD; the time, they were dismissed as behavioral problems or "attitude iss ues."[25]

Others were discharged with no medals, no debriefings, no help, no recognition of the burdens they had carried.

Even in death, inequality endured.

GRS personnel noticed that officers' bodies were often prioritized for identification and repatriation. Troops from poor or immigrant families might remain "unknown" longer, their paperwork delayed or mishandled.[26]

One Black GRS soldier reportedly escorted the body of a white officer to Iowa, then returned to base only to learn that his own cousin had died the same week, with no escort and no proper flag.[27]

These men handled the aftermath of war with grace, even as their own grief went unspoken.

They were never in the victory photos. Their stories were not televised. They served in silence.

And many of them never stopped carrying what they saw—because no one ever asked.

Interlude: No Medals for the Men Who Dug

"They said we served with honor. But they never said it to us."

In 1971, a corporal from the Quartermaster Corps returned to Oakland, California. He had spent twelve months in Vietnam—not as a rifleman, but as a GRS worker stationed at a mortuary site outside Da Nang. He had handled over 300 sets of remains.

When he arrived home, no one was there to meet him. No handshake, no flag, no recognition. Just the sound of his boots on concrete and the weight of a duffel bag filled with uniforms that still smelled like formaldehyde.

At a bar later that week, someone called him a baby killer.

"I didn't kill anyone," he said. *"I buried them."*

But the insult stayed.

He packed his uniform away. Never wore it again.

And he never told anyone what his job had been—not even his own mother.

The War at Home: Mourning Without Meaning

"They didn't want our bodies, and they didn't want our stories. We came back with caskets and ghosts."—GRS Veteran, Vietnam War

For the men of the Graves Registration Service, the most disorienting part of Vietnam wasn't the jungle. It was home.

In previous conflicts, especially World War II, the return of the war dead was surrounded by public ritual—flag-draped caskets, military salutes, and a shared language of sacrifice. The homecoming was somber but dignified. Collective. Reverent.

By contrast, Vietnam offered little more than isolation.

The war was unpopular. The country was divided. The mourning—if it happened at all—was private and often poisoned by sham e.[28]

Return to Silence

As caskets arrived at tarmacs and train stations across the United States, some families met them in silence. Others were heckled by protestors who saw every soldier—living or dead—as a symbol of political betrayal.[29]

For the men of the GRS who had escorted the bodies, documented the deaths, and processed the last physical memories of lost soldiers, the reception was often chillingly blank. There were no ceremonies for them. No thank-you notes. No space to share what they had seen.

One mortuary technician remembered flying into LAX with the body of a nineteen-year-old private. His instructions were clear: stand vigil at the funeral, return to base. The family never made eye contact. The minister never mentioned his presence. The soldier sat alone at the rear of the chapel, his uniform pressed, his grief invisible.[30]

Mourning in the Absence of Meaning

Even the families of the dead often received vague answers. The speed of notifications varied wildly. Some were told their sons had died in "noncombat" conditions without explanation. Others received the wrong remains. Some received only a tag labeled "Unknown" and a government-issued letter written in bureaucratic condolence.[31]

The public, by then, had grown tired of war.

The government, wary of backlash, limited media coverage of body returns. No televised funerals, crowds at train stations, or solemn processions.

The war's most sacred duty—returning the fallen—was now a logistical burden handled as quietly as possible. And the men who did that work carried their grief into a vacuum.

No Place to Lay It Down

Unlike other military roles, the GRS offered no brotherhood. The nature of the job—quiet, grim, disconnected from direct combat—meant that few units maintained ties after deployment. There were no reunions. No monuments. No collective memory.

They had handled the deaths. But when they returned home, there was no system to help them process their own feelings.

Many described their transition as a second exile:

- No debriefings
- No mental health services
- No formal acknowledgment
- No vocabulary for what they carried

"I didn't die," one veteran said, "but I came home like a ghost."[32]

Escorting the Dead in a Country That Didn't Want to See

One of the GRS's final roles in Vietnam was escort duty—personally accompanying remains home for burial. The job was emotionally fraught. Some families were grateful. Others refused to speak. A few asked, in quiet voices: *"What really happened?"*

There was often nothing to say.

The casket arrived. The flag was folded. The papers were signed.

And then the escort returned to base, caught another flight, or wandered into civilian life alone, changed, unmoored.

"I buried people for a living," one said. "But I never learned how to mourn."

The Weight of Unspoken Grief

Many GRS veterans never spoke of their jobs—not to families, not to therapists, not to fellow veterans. Their trauma didn't bleed outward. Instead, it simmered.

Symptoms we now recognize as moral injury and post-traumatic stress disorder (PTSD) were common: nightmares, isolation, addiction, depression.[33] Some avoided cemeteries for the rest of their lives. Others, paradoxically, found work as funeral directors or coroners, as if unable to separate themselves from death once they'd been made its steward.

There were no parades. There were no open conversations. There was only the long, quiet question: *what did we bring home?*

The GRS brought home the bodies.

But no one brought them home.

The Red Cross and the Work of Telling

"We were the voice between the telegram and the kitchen table."

In Korea and Vietnam, the American Red Cross continued the work it had begun in earlier conflicts—but now under circumstances that tested every system it had built.

Casualty notification had become more formalized since World War II. The military delivered official word of death through uniformed officers and chaplains. But the Red Cross remained essential to what came after: the questions, the confusion, the desperate need for information that official channels could not (or would not) provide.[34]

Red Cross workers staffed emergency communication centers at bases across Southeast Asia and at facilities throughout the United States. When a soldier was wounded, they helped transmit messages to families. When a soldier died, they assisted with the paperwork that connected a name in a military file to a mother's grief. They did not replace the notification system, but they humanized it.[35]

For families of the missing, the Red Cross became a lifeline. When official channels offered only silence—"Missing in Action," "Status Undetermined"—families turned to the Red Cross for answers. Caseworkers searched records, contacted military authorities, and provided what information could be found. Often, there was nothing to report. But the act of searching, of being heard, of having someone on the other end of the phone who understood—that mattered.[36]

Red Cross volunteers also supported families over the long years of waiting. They connected MIA families with one another, facilitated support groups, and advocated for government accountability. When remains were finally identified, sometimes decades later, Red Cross workers helped families navigate the process of burial, benefits, and closure.[37]

The organization's Restoring Family Links program, which traced its origins to Civil War-era tracing work and World War I prisoner-of-war communication, continued to evolve. In Vietnam, the Red Cross worked with international partners to facilitate communication between prisoners of war and their families when such communication was possible. For many families, a Red Cross message was the only proof that their loved one was still alive.[38]

One Red Cross worker stationed in Saigon described her role:

"We were the in-between. The military had its protocols. The families had their grief. We stood in the middle and tried to make them understand each other. Sometimes we succeeded. Sometimes all we could do was listen."[39]

After the war, the Red Cross continued supporting Vietnam-era families. Caseworkers helped veterans navigate the VA system. They assisted families seeking information about remains. They provided emergency assistance to those struggling with the long-term effects of service.

The work did not end when the war did.

It never does.

For the Red Cross, Korea and Vietnam reinforced a truth that Clara Barton had understood a century earlier: that war's aftermath lasts longer than war itself, and that families need support not just in the moment of loss, but in the decades that follow.

They were not the ones who buried the dead, but they were often the ones who told the living.

In that telling, they carried a burden of their own.

STILL MISSING: THE POW/MIA LEGACY

"It's not the silence that haunts us. It's the waiting."—Daughter of a Vietnam MIA, 1989.

Some men never came home. And some families never stopped waiting.

By the time the Vietnam War ended in 1975, more than 2,600 Americans were listed as Missing in Action (MIA). These soldiers had vanished into jungles, tunnels, prison camps, and uncertainty.[40] Thousands more had passed through enemy hands, with many families unsure for months—or years—whether their loved ones were alive, dead, or somewhere in between.

The Korean War had left a similar, haunting legacy. Despite formal exchanges like Operation Glory in 1954, over 7,500 U.S. service members remain unaccounted for, their bodies unrecovered and their deaths unconfirmed.[41] Most are believed to be buried in unmarked battlefield graves or inside North Korean territory, inaccessible due to decades of diplomatic impasse.

GRIEF WITHOUT A GRAVE

For families, the absence of remains created a new form of trauma—a grief suspended. There was no casket to bury. No headstone to kneel beside or telegram offering finality. Just silence.

Many of these families became activists, forming POW/MIA support networks and pressing for government accountability. Some held onto hope for decades. Others built symbolic memorials in backyards or churches.[42] A few even refused military death benefits, clinging to the belief that their sons were still alive, somewhere, waiting to be found.

"We never had a funeral," one mother said. "Because we never saw him."[43]

The GRS and the Unclosable Case

For Graves Registration Service personnel trained to retrieve, identify, and return, the MIA designation posed a terrible paradox: no recovery, no closure.

They could not dig, tag, or file. There was no ledger to complete. Instead, they processed absence. They marked files "Missing" and sent them up the chain. Some handled empty caskets for symbolic burial ceremonies. Others quietly filed death reports that read, in essence, *we don't know.*

And yet, they did not stop searching.

In the years following both Korea and Vietnam, GRS workers, forensic teams, and recovery specialists continued to pursue lost names. Bones were matched with dental charts decades later. Personal effects—dog tags, wedding rings, rosaries—surfaced in excavation sites or in diplomatic exchanges.

But for each success, hundreds remained unresolved.

The National Promise: You Are Not Forgotten

In 1971, the now-iconic POW/MIA flag was first flown, a stark black banner bearing a silhouette and a slogan: "You Are Not Forgotten."[44] It remains the only flag other than the U.S. national flag authorized to fly over federal buildings, a permanent reminder of unresolved loss.

In 1992, the Department of Defense created the Joint POW/MIA Accounting Command (JPAC) to locate and identify remains from past conflicts. It has since evolved into the Defense POW/MIA Accounting Agency (DPAA). Today, these multidisciplinary teams of archaeologists, forensic anthropologists, historians, and military liaisons continue the work once initiated by the GRS.

They visit former battlefields, interview witnesses, exhume sites across Southeast Asia and the Korean Peninsula, and send fragments to labs in Hawaii and Delaware. And when a name is finally matched to a bone, they send word to families that have waited for fifty, sixty, sometimes seventy years.

In one such case, a soldier listed as missing in 1951 was buried with full military honors in 2019. His sister, now in her eighties, placed a photograph at the grave: "We never stopped hoping."[45]

The Work That Never Ends

But for many others, the hope has thinned. Bones remain in foreign soil. War files remain sealed. Geopolitical tension continues to delay access. And each passing year means another mother, brother, or daughter dies without answers.

In cemeteries across the world, thousands of gravestones still bear one line:

"Known unto God."

These are the placeholders for the unrecovered, the unrecorded, the unnamed.

For GRS men—defined by their sacred duty to restore identity—the unknown was the one enemy they could not defeat. They could dig. They could tag. They could comfort the living.

But when the name was missing. When the face was gone. When the file read "presumed dead."

There was only the waiting. The ache. The unfinishable work.

Interlude: The Empty Chair

"We set his place at the table every Christmas. Even now. Even though we know."

The chair sat untouched for decades.

Each year, the family laid a setting: fork, knife, napkin, glass. Sometimes with his photo. Sometimes with a letter they wrote but never sent. Sometimes with nothing at all.

He was nineteen when he disappeared outside Khe Sanh and listed as Missing in Action. No remains were recovered. No definitive word was sent. Just a knock on the door.

Then paperwork.

Then silence.

The mother died before he was found. The father stopped talking about him. But the sister? She never stopped hoping.

"It's not about him coming back," she once said. "It's about knowing he's not forgotten."

And so, the chair remained.

Empty.

But not abandoned.

Spotlight: Da Nang Mortuary

Between 1967 and 1973, the U.S. Army mortuary at Da Nang processed thousands of fallen service members. Staffed by both military and civilian morticians, it was a place of precision and pain.

Air-conditioned and heavily sanitized, the facility worked around the clock, often handling up to twenty bodies per day.

Behind each body, there was a chart, a dental record, a tag. But also a story.

Morticians, including a number of African American specialists, worked in silence—performing autopsies, cleaning wounds, restoring features, and preparing remains for escort. Many developed symptoms now recognized as PTSD, but few ever received care.

It was called "the final stop before home." But for many, it was where the war truly entered their bones.

Artifact Highlight: The MIA Bracelet

Beginning in the 1970s, Americans began wearing aluminum bracelets engraved with the names of prisoners of war or missing service members. Worn until the soldier returned or their remains were found, the bracelet became more than jewelry—it was a personal vigil.

Many bracelets were never taken off. Some were buried with the wearer. Others were placed at the Vietnam Veterans Memorial Wall in Washington, D.C., where families and strangers alike still leave them, wrapped in tears and memory.

The bracelets said what the nation wouldn't: "We will not forget."

From Recovering to Accounting: The Birth of Forensic Identification

"We stopped just burying the dead. We started solving them."

The wars in Korea and Vietnam did more than produce casualties. They produced a new category of loss: the long-term missing.

And from that loss emerged a transformation in how the military approached its dead, a shift from burial to identification, from interment to investigation.

In 1976, the U.S. Army established the Central Identification Laboratory, Hawaii (CILHI) at Hickam Air Force Base.[46] Its mission was unprecedented: to recover, identify, and repatriate the remains of American service members lost in Southeast Asia and earlier conflicts. For the first time, the military had created a permanent institution dedicated not to burying the dead, but to finding them.

CILHI brought together forensic anthropologists, archaeologists, odontologists, and military investigators into a single command. They developed techniques that would revolutionize the field: DNA extraction from degraded remains, isotopic analysis to determine geographic origin, facial reconstruction from fragmentary skulls.[47] What had once been the work of gravediggers became the work of scientists.

The laboratory's teams deployed to former battlefields across Vietnam, Laos, and Cambodia. They conducted excavations at crash sites, burial locations, and areas where American personnel had last been seen. They worked with foreign governments—sometimes cooperatively, sometimes in diplomatic limbo—to access sites that had been closed for decades.[48]

The work was painstaking. A single identification might require years of analysis. A bone fragment recovered from a jungle hillside would be cleaned, measured, photographed, and compared against dental records, medical files, and DNA samples provided by families. Only when the evidence was conclusive would a name be assigned.[49]

For families who had waited decades, the call from CILHI was both ending and beginning. Their loved one had been found. The uncertainty was over. But now came the burial they had postponed for twenty, thirty, forty years—a funeral for someone who had died before some mourners were born.

CILHI also worked on Korean War cases. Though diplomatic obstacles prevented access to most of North Korea, the laboratory analyzed remains from Operation Glory and subsequent exchanges. Each identification closed a file that had been open since the 1950s.[50]

In 2003, CILHI merged with the Joint POW/MIA Accounting Command (JPAC) to expand recovery operations. In 2015, both were consolidated into the Defense POW/MIA Accounting Agency (DPAA), the organization that continues this work today.[51]

But the roots of that work lie in the aftermath of Vietnam, in the recognition that some dead cannot simply be buried, that some families cannot simply be told to grieve, that some names demand to be recovered no matter how long it takes.

The Graves Registration Service had been built to bury the dead.

CILHI was built to find them.

The shift marked a transformation in how America understood its obligations. The promise was no longer just "we will bury you with honor." It became: "We will find you. We will name you. We will bring you home."

That promise remains open.

Thousands of names are still waiting.

Known Unto Their Brothers

The terrain changed. The politics shifted. The uniforms were updated. But the grief remained the same.

From Korea's frozen ridges to Vietnam's humid fields, the men of the Graves Registration Service and Mortuary Affairs did what few could imagine or endure. They bore not rifles but stretchers. Not combat gear but tagging kits. Not slogans but silence.

They retrieved the forgotten. They honored the shattered. They gave names to the lost, even when the war itself no longer cared to remember them.

And yet.

In a filing cabinet in Hawaii, there is a folder. Inside is a name, a service number, and a grid coordinate from 1952. The grid points to a hillside in North Korea that no American has visited in seventy years. The file is marked "Open."[52]

In a cemetery in Arlington, a headstone reads "Unknown." Beneath it lie remains recovered from Vietnam in 1984, identified only as American, male, approximately twenty years old. His family—if he has one—has never been found.[53]

In a house in Ohio, a woman in her eighties keeps a photograph on her mantle. Her brother was nineteen when he disappeared near the Chosin Reservoir. She has written to every agency, every archive, every office that might hold answers. The replies are always the same: "Investigation ongoing. No new information at this time."

She still writes.

In 2023, a team from the Defense POW/MIA Accounting Agency excavated a crash site in Laos. They recovered bone fragments, a dog tag, and a wedding ring. The identification process will take years. The family has been notified that remains have been found. They have not been told whose.[54]

The work continues.

Forensic anthropologists analyze fragments in laboratories. Diplomats negotiate access to burial sites. Families send DNA samples and wait for the phone to ring. Files that were opened in the 1950s and the 1970s remain open. The missing do not become less missing with time.

They only become older.

And the caretakers? Some died carrying what they saw. Some are still alive, still silent, still waiting for someone to ask what it cost.

A former mortuary specialist, now in his seventies, lives alone in a small town in Oregon. He has never told his children what he did in Vietnam. He has never visited the Wall. He has never spoken to a counselor or joined a veterans' group.

But every November, he drives to the local cemetery and stands for an hour beside a grave that is not his own. A friend from the war. Someone he processed at Tan Son Nhut in 1969.

He doesn't pray. He doesn't speak. He just stands there.

Because someone should.

The wars in Korea and Vietnam asked: *what do we owe the dead we cannot find?*

The caretakers gave their answer: *everything. We owe them everything.*

They searched. They documented. They waited. They carried files and memories and silence for decades.

And they are still waiting.

Because the work is not finished.

Because the names are not all recovered.

Because somewhere, in a jungle or a frozen hillside or a forgotten corner of a forgotten war, someone is still missing.

And someone, somewhere, is still looking.

Wars Without Closure, Burials Without Witness

Unlike the sweeping ceremonies of World War II, Korea and Vietnam offered little in the way of national mourning. One war was paused but never ended. The other closed with a televised helicopter lift and years of cultural amnesia.

For the GRS, it meant that their labor—the sacred act of preparing the dead for burial—occurred in political limbo. Their work didn't change, but the nation's willingness to face it did.

They flew bodies home to airports no one met. They stood at funerals where families trembled in confusion, grief, or shame. They folded uniforms and sealed footlockers filled with blood-stained photographs, medals, and final letters.

And then they returned to base and did it again.

Caretakers in a Country That Turned Away

The public rarely knew them. The government rarely acknowledged them. Their roles were classified as "support." Their trauma labeled "behavioral."

But they carried war in their hands—hands that unzipped bags, tied tags, combed hair on lifeless heads, and closed the eyes of boys who died far from home.

They were asked to do the impossible: to bring dignity into chaos. And when they returned, there was no ceremony. No counseling. No memorial. Only silence.

Known Unto Us

They buried strangers. They buried brothers. And in time, they buried pieces of themselves.

What they gave us was not just graves; it was certainty. It was the quiet knowledge that someone had seen our dead not as numbers, but as names.

They were the last to hold them. The last to speak for them. The last to say: *"You mattered."*

And so this chapter, and this book, names them too.

Not only those who fell, but those who carried them. Not only those we memorialize, but those we have overlooked. Not only those who died, but those who survived and were never asked what that survival cost.

Let Us Remember What They Carried

These were not warriors in the traditional sense. They were grief-bearers, assigned to carry death through wars that never offered them healing. They did not return with medals—they returned with memories. And they did not tell stories because no one wanted to hear them. But now, we listen. Because in honoring them, we recover something else we lost: The belief that even in war, every life matters. And so does the hand that lays it down with care.

Reflection

Once a month, I visit veterans at the VA hospital. They are being treated for psychological wounds—some from recent deployments, some from wars that ended decades ago. The trauma doesn't care about calendars. It lives in the body, in the memory, in the dreams that won't stop coming. We do art with them. Simple things like coloring, painting, and crafts that keep the hands busy while the mind finds space to breathe. Sometimes they talk about what they've seen. Sometimes they just sit in silence, grateful for company that doesn't demand anything from them.

And sometimes we bring stars.

These are stars cut from retired American flags—flags that have flown over government buildings, been used in ceremonies, and reached the end of their service life. Rather than dispose of them, volunteers cut out the stars and give them to veterans. The reaction is always the same. These men and women—trained to be tough, stoic, and controlled—take the small fabric star and weep. They hold it close. They carry it with them. Something about that worn piece of flag, something about being seen and honored, breaks through whatever walls they've built.

One veteran told me he keeps his star in his pocket at all times. "It's like carrying the country with me," he said. "The good parts. The parts worth remembering." Witnessing these moments has taught me something this chapter tries to convey: that trauma does not heal itself.

Silence is not strength. What is honored in public often fractures in private. The caretakers—those who carried the dead, who witnessed what war leaves behind—need to be carried themselves.

The flag stars are small gestures. But small gestures, repeated faithfully, can hold enormous weight.

Someone must witness the witnesses.

That is the work I try to do.

PART III

BEYOND THE BATTLEFIELD

"War doesn't end when the fighting stops. It ends when the dead are at rest—and the living are at peace."

Up to this point, we have followed those who buried the dead during war. But what happens after?

This part of the book shifts focus from trenches and transport planes to the spaces where memory is made and carried forward. These chapters explore the long aftermath of war death, tracing its impact through forensic science, ritual, cultural identity, and the unseen labor of those who continue to tend the fallen long after the last gun is fired. It is here that we meet the death workers of peace—the archaeologists who unearth forgotten bones, the scientists who name the nameless, the Indigenous communities who carry their warrior traditions into modern remembrance. We explore what it means to bury an enemy and why ritual matters even in opposition.

Here, the focus is on care and the universal human need to make meaning of death.

This part of the book invites the reader to move inward into anthropology, identity, and the shared fragility that binds all who mourn.

It asks how we remember. How we honor. How we heal. And what happens when we do not. Because even after the battlefield is cleared, memory remains. And someone must carry it.

REFLECTION

The funeral was precise. The uniforms were pressed, the shoes shined, the movements practiced and exact, as if discipline itself could hold the moment steady.

My grandfather's casket rested beneath a folded flag. A chaplain spoke words I had heard before, meant to comfort and to confine death into something orderly and understood.

Then the honor guard took their positions.

I knew what was coming. I had seen it on television, at ceremonies, from a distance. I knew the salute was meant to honor service, to mark sacrifice, to close a life with dignity.

I was not prepared for the sound.

The first shot cracked the air and went straight through me. Not metaphorically—physically. My chest seized. My breath caught. The ground seemed to jump beneath my feet.

Then came the second.

And the third.

Each report landed with the same force, a measured violence delivered in perfect time. Twenty-one controlled explosions, each one echoing longer than the last, reverberating through bone and muscle and memory.

I stood frozen, my body reacting before my mind could catch up.

This was honor. This was ceremony. And yet my hands trembled, my pulse raced, and something deep in me recoiled—not from fear, but from recognition.

This was the sound of war, stripped of chaos and given order. The sound of rifles firing not to kill, but to remember killing.

I realized then that ritual does not erase violence. It contains it.

The salute was meant for my grandfather, but it did not belong to him alone. It carried every battlefield he'd survived, every body he'd seen, every moment he'd never spoken aloud. It carried the war he'd brought home and the war he'd kept buried.

When the final shot faded, the silence that followed felt heavier than the sound itself. A silence dense with everything the ceremony could not name.

In that moment, I understood something this book has been circling ever since.

Even when war ends, the body remembers.

Even when violence is ritualized, it still shakes us to our bones.

Intermission

Between War and Remembrance

"There was no more blood. Only memory. And we were the ones left to carry it."—Vietnam War Graves Registration Clerk

The guns fell silent. The bodies were buried. But the war did not end.

It lingered in the eyes of the men who had filled the graves. In the hands of those who folded uniforms, labeled boxes, and wrote the names again and again until their knuckles ached. In the quiet rituals performed in empty rooms.

The men and women who tended the fallen did not return home unchanged. They returned carrying something invisible, something heavy, something sacred.

The world moved on. They could not.

This intermission stands as a pause between the physical act of burial and the psychological, cultural, and spiritual labor that follows. It is a breath held between war and the remembrance that must come next.

Because those who buried the dead did more than close graves. They opened a new kind of mourning.

There is a kind of death that does not happen on the battlefield. It happens later—in the quiet. In the remembering.

For those who carried the dead, the trauma of war did not end with survival; it began with what they were asked to witness, hold, and bury. These were not just soldiers but witnesses of war's aftermath, tasked with preserving dignity where there was none and often left to carry their grief alone.

Chapter Seven

The Burden of Memory: Trauma and the Living Dead

This chapter explores what happens to those who live with war's dead—not just physically, but emotionally, spiritually, and psychologically. It follows the men and women who zipped the bags, dug the graves, logged the names, and went home to lives that no longer felt whole. Some suffered silently. Some fell apart. Others became caretakers of memory in different forms, haunted but holding on.

Here, we examine the concept of moral injury, the long-ignored trauma of noncombat roles, and the crushing weight of silence endured by medics, morticians, chaplains, and recovery teams. We explore the spaces, both institutional and internal, where grief was buried alongside the dead.

And we ask a difficult question: what happens when no one mourns the mourner?

This chapter does not offer resolution.It offers recognition.Of pain. Of silence. Of the cost of carrying the dead—and surviving it.

Silent Echoes: The Invisible Burden of the Gravediggers

"We didn't die, but we died in different ways."—GRS Veteran,

post-WWII

They were not combat soldiers. They did not storm beaches or charge enemy lines. But they were never far from the front.

The men assigned to the care of the dead were surrounded by reminders of mortality. Day after day, war handed them broken bodies, pieces of uniforms, and names on dog tags. Their job was not just to bury the dead—it was to carry them. Through the mud, through the paperwork, through memory. And that weight never lifted.

Long before terms like post-traumatic stress disorder (PTSD) entered the public vocabulary, these men were living its symptoms: insomnia, nightmares, flashbacks, numbness, and panic attacks. They returned from war but could not return to peace. They woke at night seeing faces and drank to forget smells. They avoided touch because their hands still remembered what they had held.

In World War I, such trauma was often misdiagnosed as shell shock, a condition believed to stem from physical damage to the brain due to artillery blasts. But for those in the Graves Registration Service who rarely faced incoming fire, the damage came from relentless exposure to death rather than explosive force. They saw what war did to the body—again and again—without adrenaline, without distraction, and without recognition.[1]

By World War II, the psychiatric field had begun to develop frameworks like combat fatigue, yet burial workers remained largely overlooked.[2] They were not seen as warriors—they were seen as logistics, support staff, noncombatants. Their silence was expected, even institutionalized.

In Korea and Vietnam, the silence deepened. Veterans of these wars came home to a country that had little interest in hearing what they had endured. Few mentioned they had served in Mortuary Affairs. Even fewer admitted what that service had done to their minds. "You buried the dead?" people asked. And then:

"Let's not talk about that."

But the mind cannot obey silence forever.

What could not be spoken found other forms. Some men became nauseated at the smell of soil. Others compulsively cleaned themselves or their homes, trying to wash away what could not be scrubbed from memory. Others withdrew from family life entirely, fearful they might transmit the darkness they carried. They became watchful fathers, distant husbands, quiet men who couldn't explain why certain sounds or textures shattered them.

Modern clinicians have come to recognize this not only as PTSD, but as moral injury—a psychic wound inflicted when individuals are forced to violate their own moral or ethical codes, even when following orders.[3] These were not killers, but they were witnesses to killing. Though their task was not violence, it was soaked in violence.

And it was lonely.

Burial men were often separated from the rest of their units. Other soldiers kept their distance—not out of disdain, but out of fear. "If I see you," one infantryman told a GRS private in Vietnam, "it means someone's not going home."

And yet, every soldier did see them. Every day. In those brief encounters, the living glimpsed their own mortality.The GRS men carried that reflection—for all of them.

The cost of caring for the dead was not only psychological but moral, spiritual, and permanent.

The Cost of Caring: Psychological Toll of Burial Work

"You don't see the blood on our uniforms, but it's there. Always there."
—Korean War Recovery Specialist

They did not fire weapons.They filled out forms.They zipped the bags.And they bore a grief that had no ceremony, no language, no medal.

The psychological toll on burial and recovery teams was both profound and largely unacknowledged. Unlike combatants whose trauma was often validated through visible wounds or battle narratives, the men of Mortuary Affairs and Graves Registration endured what scholars now identify as vicarious trauma—a condition stemming from repeated exposure to the suffering or death of others.[4]

Their proximity to the dead was unrelenting. In World War II alone, GRS units handled over 280,000 American remains in the European Theater, sometimes processing hundreds of bodies a week.[5] In the Pacific, the conditions were often worse—high heat, rapid decomposition, and the brutal recovery of remains from jungles, mountainsides, and ocean waters. The psychological effects of this work were compounded by the physical environment, often leaving personnel ill, exhausted, and emotionally raw.

By the Vietnam War, documentation shows that even among support services, mortuary units had some of the highest rates of alcohol abuse and disciplinary transfer, often linked to unaddressed trauma.[6] These men weren't suffering because of what they did. They suffered because of what they had to carry, mentally and emotionally, with no outlet for release.

Many of them were young—barely twenty—and had never seen death before being asked to collect, clean, and catalog it daily. Some were tasked with recovering the remains of people their own age. Others were handed charred, fragmented, or partial remains, and ordered to render them recognizable. The emotional cost was staggering. The act of touching the dead, of washing blood from uniforms, of removing wedding bands and transcribing personal effects demanded an impossible emotional duality: compassion without collapse.

Those who stayed in the work long enough developed what one Vietnam-era specialist called a "split mind," a compartmentalization strategy used to survive.[7]

On the job, they moved efficiently, precisely, even stoically. But off duty, many unraveled. Some were haunted by dreams. Others lost their appetites or developed obsessive habits. A few left the military altogether, unable to reconcile the silence expected of them with the grief exploding inside.

Unlike front-line soldiers who often bonded over shared trauma, GRS teams were isolated. Their job was taboo, and their presence reminded others of mortality. This isolation only deepened their pain. Medics, chaplains, and even psychiatric officers rarely visited the recovery tents or mortuary zones, further reinforcing the idea that this trauma was not "real soldiering."

Modern psychologists now understand this reaction as part of what's known as secondary traumatic stress. In a civilian context, it's seen in coroners, nurses, or forensic investigators. But in war, it was most often borne by the burial men.

And they bore it in silence.

One GRS veteran from World War II later told an interviewer, "I knew every soldier I processed had a mother. A name. A toothbrush.

And I had to forget all of that, or I'd fall apart."

They didn't forget.They carried it.And that carrying cost them dearly.

CALLED TO THE DEAD: THE PSYCHOLOGICAL SELECTION OF MORTUARY PERSONNEL

"You had to be the kind of person who could look death in the face and not look away.That didn't mean it didn't hurt. It meant you did it anyway."—Army Mortuary Affairs NCO, Iraq

Not everyone can do this work.

The duty of caring for the dead in war—retrieving, cleaning, identifying, transporting, and burying—requires more than physical resilience. It demands an emotional and spiritual fortitude that few can sustain. Some soldiers were assigned to Mortuary Affairs or Graves Registration without choice; others, more rarely, volunteered. Regardless of how they arrived, all were transformed.

By the post-Vietnam era, the U.S. military recognized that not all personnel were suited for this emotionally grueling task. The Army Mortuary Affairs Center, restructured during the Gulf War era, began incorporating elements of psychological screening into training programs.[8] Recruits were evaluated not only for physical capability but also for their ability to maintain emotional stability under prolonged stress and exposure to corpses.

But even before formal screening processes, certain qualities set these individuals apart. Interviews with Mortuary Affairs veterans across multiple eras—from Korea to Iraq—consistently reveal shared traits:

- A high threshold for emotional endurance

- An instinct for meticulous detail amid overwhelming grief
- A drive to bring dignity to chaos
- A profound sense of duty, even when unrecognized

These were not hard-hearted men. They were not cold. Many describe themselves as deeply sensitive—precisely the kind of people who felt the magnitude of what they were doing. They had what some military chaplains call the "quiet call," a vocation rooted not in glory, but in compassion.[9]

Some were drawn to the dead due to previous civilian experience in funeral homes, coroner's offices, or small-town cemetery work. Others entered the role after exposure to loss or trauma in their own lives. For some, the work was about control—in a world fractured by war, handling the dead offered one last place where order could prevail.

And yet, the cost was never far behind.

Despite the shared traits and training, many Mortuary Affairs personnel eventually suffered symptoms of vicarious PTSD, complicated grief, or depressive episodes.[10] While their job demanded detachment, the reality was often different. Soldiers do not process corpses in the abstract. They tag the belongings of the newly dead, recognize names from shared mess halls, and sometimes recover friends.

In Iraq and Afghanistan, units that handled the remains of service members killed by IEDs were among the most vulnerable to trauma.[11] The condition of the bodies, often mutilated or burned, left emotional scars that no training could prevent. One GRS veteran of Operation Iraqi Freedom later shared: "I learned how to be calm. I learned how to smile. But I couldn't listen to country music again—too many lyrics about home, and that's where the dog tags went."

The military attempted to mitigate the psychological toll through debriefings and chaplain support. But many soldiers found little comfort in clinical approaches. What they needed was often simpler, yet far harder to find: someone willing to sit with their grief, to witness it without turning away.

Because ultimately, those who were called to the dead didn't just serve the fallen.They bore witness to the unbearable—so that families wouldn't have to.They saw what others could not.And they stayed, when others walked away.

The Wider Circle of Witness: Chaplains, Medics, and Those Who Carried Alongside

"I wasn't Graves Registration. But I stood at the same graves."—Army Chaplain, Vietnam, 1969

The men of Mortuary Affairs and Graves Registration bore the most sustained exposure to the dead. But they were not alone in carrying the burden. Around them moved a wider circle of witnesses—chaplains, medics, nurses, clerks—each touching death in ways that left marks the military rarely acknowledged.

Chaplains occupied a unique position. They were expected to provide spiritual comfort to the living while officiating over the dead, sometimes conducting multiple burial services in a single day. During World War II, chaplains accompanied GRS teams to temporary cemeteries, offering prayers over bodies that had no names and families that would never hear those prayers.[12] In Vietnam, chaplains served at mortuary facilities, blessing remains before transport and counseling mortuary workers who had no other outlet for their grief.[13]

The burden was theological as well as emotional. Chaplains were asked to make meaning of deaths that often seemed meaningless. They prayed over the bodies of nineteen-year-olds killed by friendly fire, civilians caught in airstrikes, and enemy soldiers whose names they would never know. One Army chaplain, interviewed decades later, described the work as "standing at the threshold between the living and the dead, trying to say something that mattered to both."[14]

Many developed what clinicians now recognize as compassion fatigue, the emotional exhaustion that comes from sustained empathic engagement with suffering.[15] But during their service, few received support. They were expected to support others. The idea that the chaplain himself might need care was rarely entertained.

Medics and corpsmen carried a different wound. Their job was to save lives, and when they couldn't, they were often the first to witness death. Unlike GRS personnel who arrived after death had occurred, medics held soldiers as they died. They performed CPR on men already gone. They made impossible decisions about who could be saved and who could not.[16]

The transition from trying to save to accepting death was a daily occurrence in combat zones. Medics who failed to save patients often experienced profound guilt, even when nothing could have been done. This guilt compounded over time. One Navy corpsman who served in Vietnam later said: "You remember the ones you lost. Not the ones you saved. It's backward, but that's how the mind works."[17]

After the patient died, medics sometimes assisted with remains—helping to bag bodies, record information, or transport the dead to collection points. This dual role as both healer and handler of the dead created a psychological burden that was neither fully medical nor fully mortuary. It fell between categories, and so did the men who carried it.

Nurses, particularly those serving in combat support hospitals, witnessed death with a frequency that their civilian counterparts rarely approached. Army and Navy nurses in World War II, Korea, and Vietnam often worked in facilities where patients died despite all efforts, where mass casualty events filled wards with the dying, and where the line between treatment and comfort care blurred hourly.[18]

Some nurses assisted with post-mortem procedures, washing bodies, inventorying personal effects, and preparing remains for transport. Others wrote letters to families, translating the clinical reality of death into words a mother could bear to read.

This work was rarely acknowledged as part of nursing duties, but it was performed nonetheless.[19]

Clerks and typists carried a burden that was administrative but no less real. The women and men who typed death notifications, casualty reports, and condolence letters processed thousands of names. Each name was a person. Each form was a family's world ending. One WAC clerk who served in the European Theater during World War II later recalled: "My fingers knew the shape of 'killed in action' better than my own name. I typed it in my sleep."[20]

These clerks were rarely considered part of the deathcare system. They were "administrative support." But they absorbed the scale of loss in ways that combat troops, focused on survival, often did not. They saw the numbers. They knew the patterns. And they carried that knowledge in silence.

The trauma of witness was not confined to those who handled bodies. It extended to everyone who touched death's bureaucracy, who stood at its threshold, who processed its paperwork, who offered its prayers.

They were all caretakers.

And they all paid a price.

Invisible Wounds: Silence, Stigma, and Shame

"You had to keep it inside. There was no place to put it, so you just... swallowed it."—GRS Veteran, Korean War

They wore no Purple Hearts. They returned with no visible wounds. And so, no one asked what they had seen.

In the cultural mythology of war, trauma is often validated through visibility. A scar, a prosthetic, a battlefield story told through medals or documentaries—these are accepted markers of sacrifice. But for the men of the Graves Registration Service, trauma hid in quieter places: in the night sweats, in the silence at dinner, in the eyes that flinched at a folded flag.

Their wounds were internal, and so the world moved on.

Most were demobilized without psychological care. They were told to return to civilian life, to find jobs, to "be grateful" that they had come back alive. But many of them didn't feel alive. They felt haunted. And worse—forgotten.[21]

In many cases, even their families didn't know what they had done. Some were listed as clerks. Others simply said they "worked logistics." One Vietnam-era GRS veteran tried to explain his nightmares to his wife. "She thought I'd been behind a desk. When I told her I handled kids... bodies... she went quiet and never brought it up again."[22]

The cultural silence around their role wasn't just personal—it was institutional. Military histories often omitted their contributions. Hollywood rarely showed their labor. Exhibits in major war museums, if they featured Mortuary Affairs tools at all, presented them without names, context, or stories.[23] Even within veterans' organizations, the men of GRS were often the quiet ones at the back of the room. They clapped at speeches, they saluted at memorials. But they rarely stepped up to the mic.

This absence of voice had consequences. It reinforced the idea that what they had seen didn't matter, that what they had suffered couldn't compare, that grief must be borne alone.

Modern psychologists describe this experience as disenfranchised grief—a form of mourning that is not acknowledged, validated, or supported by society.[24] These men grieved not only for the dead they buried, but for the selves they lost in doing so. Yet their grief found no audience, no outlet, no ritual. And in many cases, it turned inward, into shame, self-blame, or numbness.

Some carried guilt for errors made under stress—missed tags, misfiled reports, a misidentified face. Others carried the emotional residue of young bodies or civilian casualties they were ordered to process without comment. These memories replayed, again and again, like static in the background of their lives.

Even as awareness of combat PTSD began to grow in the late 20th century, GRS trauma remained largely invisible. Because they were not "on the front lines," many veterans of burial units were disqualified from treatment programs or denied support from their VA casewor kers.[25]

The criteria for psychological injury too often failed to account for moral labor.

And the men knew it.

"People think we just dug holes," one World War II veteran said in a 1990 interview. "They don't know what it's like to hold a piece of someone and try to pretend that's enough."[26]

That kind of knowing—unspoken, heavy, holy—is a wound that never fully heals.

Their pain was not born from action, but from proximity.Not from violence committed, but from violence witnessed.And when the world failed to make space for that truth, many simply buried it.

The result was a generation of silent survivors who could not explain why they avoided funerals, why they cried at taps, why they drank on anniversaries of battles they'd never fought.

And now we know why.

The wounds they bore were not etched in skin.But they were etched in memory.

Families and the Inheritance of Silence

"I grew up with a ghost. My father was alive, but part of him had never come home."—Daughter of a WWII GRS veteran, 2015

The trauma of burial work did not stay with those who performed it.

It traveled.

Into marriages. Into parenting. Into the fabric of families who lived with men they loved but could not fully reach.

The veterans of Graves Registration and Mortuary Affairs who returned from war often returned changed in ways their families could sense but not name. They were present at dinner tables but absent in conversation. They flinched at sounds their children could not hear. They avoided funerals, cemeteries, Memorial Day gatherings—the rituals that should have offered closure but instead reopened wounds.[27]

Spouses bore the weight of that withdrawal. Many learned early which topics to avoid, which questions not to ask, which silences to respect. Some marriages survived through accommodation, a careful choreography of avoidance that preserved the peace but prevented intimacy. Others did not survive at all.[28]

One wife of a Korean War GRS veteran described fifty years of marriage as, "living beside a locked room. I knew something was in there. He knew I knew. But we never opened the door."[29]

Children experienced a different kind of loss. They grew up sensing absence within presence—a father who was physically there but emotionally unreachable. Many described their fathers as loving but remote, capable of tenderness but incapable of explanation. The war was never discussed. The questions were never answered. The silence became a third presence in the household.[30]

Psychologists now understand this phenomenon through the concept of intergenerational trauma—the transmission of psychological wounds across generations, often without explicit commun ication.[31] Children absorb their parents' emotional states even when the content remains hidden. They learn anxiety, hypervigilance, and emotional suppression not through words, but through watching.

Literary scholar Marianne Hirsch has termed this phenomenon "postmemory"—the way traumatic experiences are transmitted to those who did not live through them but who grew up in their shad ow.[32]

Children of Holocaust survivors, of combat veterans, and of those who witnessed atrocity all may carry memory that is not their own but that shapes their lives nonetheless.

The children of Graves Registration veterans often describe a particular inheritance: the sense that death was always nearby, that grief was too large to be spoken, that their fathers had seen things that could not be shared. Some became caretakers themselves—funeral directors, hospice workers, chaplains—as if drawn to complete what their fathers had left unfinished. Others avoided anything connected to death, as if proximity might unlock whatever their fathers had sealed away.[33]

For many families, understanding came only after death. When fathers passed and attics were cleaned, children found the shrapnel boxes, the footlockers and folders containing field maps, carbon copies, photographs never shown, and letters never sent. These artifacts offered posthumous explanations for lifetimes of distance.

One daughter, interviewed for an oral history project, described opening her father's footlocker after his funeral: "I finally understood why he couldn't talk about it. And I finally understood why he couldn't talk to me."[34]

The silence was not cruelty. It was containment.

But containment has costs. And the families paid them too.

Today, recognition of caretaker trauma has begun to include family support, including programs that help spouses and children understand what their loved ones experienced and support groups that offer community to those who lived with silence. The VA has expanded its family services, and veteran organizations have created space for adult children to share stories their parents never told.[35]

But for many families, understanding came too late. Fathers died without speaking. Children inherited questions that will never be answered.

The war did not end when the veterans came home. It moved into their homes.

And it lived there, quietly, for generations.

Haunted by the Ledger: The Rituals that Linger

"I don't remember birthdays anymore. But I remember every name I ever wrote down."—WWII GRS clerk, oral history, 1984

The work was procedural. But the pain was not.

For the men of the Graves Registration Service, the daily routines of their task—tag, record, bury, repeat—became more than duties. They became rituals. And like any deeply ingrained ritual, they left lasting marks on the psyche.

Every recovery followed a process. Unzip the bag.Check the tag. Record the serial number. Note the condition of the remains. Assign a temporary grave. Map the site. Log the file. Repeat.

In the chaos of war, this routine offered a kind of order—one of the few things these men could control. But repetition does not equal detachment. For many, the details embedded themselves into memory. A name that sounded familiar. A photo tucked in a helmet. A note folded inside a uniform pocket. These were the fragments that lingered, anchors of grief wrapped in bureaucracy.

Some GRS soldiers kept unofficial records: small ledgers, duplicate field forms, hand-drawn grave maps. Not out of regulation, but out of reverence—or guilt. These documents were sometimes stored in footlockers or hidden in personal Bibles. Some were passed down through families. Others were burned in the backyard after nightmares returned.[36]

One World War II clerk confessed decades later that he could still recite dozens of names from a single campaign. Another described waking in the night to recheck fictional entries in his mind. The procedures never left them. They had become internalized liturgies, a catechism of grief.

In anthropology, death workers have long held sacred roles. Across cultures, the task of preparing the dead often falls to priests, midwives, or shamans, individuals trained not just in the practicalities of burial, but in the spiritual act of guiding souls across thresholds.[37] These rituals are meant to serve the dead, but also to preserve the emotional balance of the living.

For GRS personnel, there was no ritual for themselves. No cleansing rite, no mourning period. Their job was to absorb the unspeakable so others didn't have to. But without spiritual tools or cultural language, that burden had nowhere to go. It calcified.

Some veterans reported becoming obsessive organizers later in life—alphabetizing everything, over-documenting trivial tasks, or panicking when forms were misplaced.

Others abandoned paperwork altogether, unable to approach a blank form without physical discomfort.[38]

The trauma of "necessary ritual" shaped not only minds but lives.

One man refused to attend his own brother's funeral. Another declined to serve as a pallbearer—not because he was unwilling, but because the act of carrying another coffin would have undone him. Still others found themselves unconsciously "tagging" the dead in movies, car crashes, or news reports, mentally returning to wartime reflexes as if still on duty.

These weren't memories.They were *habits.*Muscle memory formed in the crucible of war, long after the danger had passed.

Perhaps the most heartbreaking legacy of these rituals was their intimacy. These men knew the dead better than anyone. They'd handled their effects. Cleaned their wounds. Seen their scars. Said their names out loud before closing the file.[39] But the connection was one-sided. There were no thank-yous. No wakes. Just silence.

They performed these rituals with reverence. And then went home to no ritual at all. The community never gathered around them, the nation never heard their names, and the burden of memory was left to rot in the corners of closets and inside dreams.

And yet, in their quiet repetitions, these men offered the world something holy:

- That grief deserves order
- That names deserve to be spoken
- That death deserves witnesses

Even if those witnesses were left alone in the end.

Bodies but No Words: The Ethics of Witnessing

"We didn't have the words for what we saw. We had forms. We had numbers. But we didn't have language."—GRS Mortuary Specialist, Vietnam War

There is a kind of trauma that comes not from being wounded—but from watching others be destroyed.And being told to move on.

The men and women of the Graves Registration Service were not passive observers. They were not noncombatants in the traditional sense. But they were not warriors either. They existed in a liminal space—summoned after violence had passed, yet still entangled in its immediate aftermath. Their job was not to stop suffering, but to document what suffering left behind.

That role carries an ethical weight.

These caretakers of the dead stood beside bombed-out villages, recovered the remains of civilians caught in artillery fire, pulled soldiers from tank wrecks, minefields, rice paddies, and collapsed bunkers. Some worked in regions where entire units had been wiped out, leaving only fragmented bodies behind. Others handled the dead after ambushes or air strikes, situations that often blurred the line between tragedy and atrocity.[40]

Sometimes they recovered children.Sometimes they knew the name on the tag before turning it over.Sometimes they had to piece together someone they'd eaten breakfast with that morning.

There were no counselors in the field. No chaplains at the morgue. No ethical framework for what they were seeing. Just standard operating procedure:

"Condition of Remains: Partial. Cause of Death: Blast injury. Personal Effects: Rosary, wallet, letter (unread)."

The military language—clean, clinical—offered no vocabulary for grief. No ritual for making meaning.[41]

And yet, the human mind does not process death like a form. It stores memory in images, sensations, smells. What GRS personnel experienced was more than technical datal; it was moral proximity to suffering, and it left an imprint deeper than the body.[42]

Many bore witness to events that they knew were wrong, even if no one around them said so—a rushed burial, a misfiled identification, a civilian marked as "collateral." And though they weren't the ones giving orders, they were the ones who had to hold the consequences. What philosopher Susan Sontag calls "the burden of the witness" applies powerfully here: *to see and say nothing is its own form of injury.*[43]

Even decades later, many veterans remained haunted not just by what they had seen, but by what they had been unable to say. The silence was not cowardice—it was containment. Speaking of it felt impossible. But so did forgetting.

This phenomenon is now studied under the umbrella of moral injury, a concept distinct from PTSD. While PTSD is rooted in fear, moral injury stems from betrayal, helplessness, or complicity in acts that violate one's deepest values.[44]

For GRS teams, moral injury often came from the ethical contradictions embedded in their role. They were tasked with dignifying death but often denied the dignity of reflection. They bore witness to horror but were not allowed to speak of it. They recorded the facts but were never permitted to grieve the story.

One Vietnam-era mortuary clerk remembered being asked to help transport the body of a young girl mistaken for an enemy combatant. "She had a ribbon in her hair," he recalled. "I kept thinking—someone packed her lunch that morning." He filed the report, wrote "Civilian," and never discussed it again for forty years.

This is not trauma born of action.It is trauma born of presence.Of being the one left standing, clipboard in hand, while others died.

These were not heroes in the mythic sense. But they were, as theologian Rita Nakashima Brock writes, "keepers of the unbearab le."[45] They carried the spiritual weight of a war they did not choose, absorbing its worst moments so others wouldn't have to.

They did not ask to be witnesses. But they did not look away.And in that gaze—silent, steady, and unthanked—there was a sacred kind of resistance.

Interlude: Beneath the Silence

"You don't see the blood on our uniforms, but it's there. Always there."

—Korean War Recovery Specialist

They came home without visible wounds, and so, no one asked what they had seen.

They folded into society like shadows, holding jobs, raising children, sitting in church pews. But in the quiet, they remembered. The names. The faces. The smell of formaldehyde and blood. The silence after a mother's sob. The dreams that wouldn't end.

There was no monument for them. No medals. Only memory.

And a grief that had no ritual, because the world never made one for those who buried the war.

Healing and Memory: Veterans Who Spoke, and Those Who Couldn't

"It took me 40 years to talk about it. And once I did, I couldn't stop."

—GRS veteran, World War II oral history

For decades, they said nothing.

Their silence was not born of indifference, but survival. To name what they had seen would be to relive it. For the men of the Graves Registration Service, words were dangerous. They threatened the fragile structures these veterans had built to keep themselves intact—habits, routines, distance.

And so most did not speak.

They carried their stories like sealed envelopes—private, heavy, undeliverable. Not to wives. Not to children. Not even to other veterans, unless they, too, had zipped the same bags and logged the same names.

This silence had consequences: children grew up sensing distance they could never explain, spouses learned when not to ask questions, and friends avoided military topics, unsure of what would provoke withdrawal or anger.

But silence is never absolute.It is a container.And eventually, some began to open it.

In the late 20th century, as public language around PTSD and moral injury expanded, a number of GRS veterans began to speak. Some were prompted by oral history projects; others by aging and the sense that time was running out. The emergence of institutions like the Library of Congress Veterans History Project and university-sponsored narrative collections gave these men a platform they had never been offered before.[46]

The stories that emerged were raw, trembling, and often redemptive. They spoke of children's shoes found in rubble. Of wedding rings recovered from crushed fingers. Of recording names that still echoed in dreams. Some wept during interviews. Others clutched their notebooks, reciting the steps of identification procedures as if reciting the Rosary. These were not performances, they were confessions—slow, solemn, sacred. For many, the act of speaking became its own kind of ritual healing.[47]

Words, after years of silence, were not just informative. They were purgative. A cleansing act.

But not all could speak.

Some veterans had buried their experiences so deeply that to exhume them would have felt like desecration. They feared what might come loose. Others had tried before only to be met with awkwardness or dismissal. "That was a long time ago," someone would say. "You should be over it by now."

And so, they swallowed it again.

Yet even these silent veterans left behind traces in the way they folded flags with aching precision. In their refusal to attend funerals. In the sudden quiet that followed war movies. In their Memorial Day rituals—small, private, haunted.

Their silence became a form of speech, an embodied language of grief without words.

Scholars now understand this phenomenon through the concept of postmemory, a term coined by Marianne Hirsch to describe the transmission of trauma across generations, especially when the trauma is unspeakable.[48] Children of these veterans often describe growing up with a "shadow parent"—someone deeply loving, yet unreachable. Some inherited the burden of that silence without ever hearing the story behind it.

The good news is that the silence is beginning to break.

In recent years, museums, military chaplain corps, and mental health organizations have begun to recognize the distinct trauma faced by noncombat military personnel, especially those in death-related roles. The U.S. Army Chaplaincy School now includes case studies on moral injury among mortuary workers.[49] Counseling centers offer writing workshops and trauma storytelling programs for veterans who served in GRS and Mortuary Affairs.

These stories are reshaping not just memory—but healing it.Not just history—but humanity.

Because, in truth, the GRS veterans never wanted monuments. They only wanted to be remembered truthfully.[50]

Now, we have the language to do so.

Whether whispered, written, recorded, or held in trembling hands, their stories are no longer lost. They are part of the collective inheritance of war's memory. And in hearing them, we extend the same dignity they gave so many others:

To be named.To be known.To be brought home.

The Red Cross and the Parallel Wound: What Disaster Response Teaches Us About War

"We weren't in combat. But we were surrounded by death."—American Red Cross Disaster Mental Health volunteer, 2005

The trauma carried by military caretakers is not unique to war. It has a parallel in civilian disaster response—and understanding that parallel illuminates both.

American Red Cross workers have long served at the intersection of catastrophe and care. They respond to hurricanes, wildfires, mass casualty events, and the quiet disasters of house fires and sudden death. They do not carry weapons. They carry blankets, paperwork, and the willingness to sit with people in their worst moments.[51]

And like military caretakers, they pay a price.

The Red Cross recognized secondary traumatic stress—the psychological impact of repeated exposure to others' trauma—as an occupational hazard long before the military acknowledged it in noncombat roles.[52] Disaster Mental Health programs, developed in the 1990s, were designed not only to support disaster survivors, but to support the responders who served them. The organization learned what military institutions were slower to accept: that caring for the traumatized is itself traumatic.

The symptoms are familiar. Disaster responders report nightmares, intrusive memories, emotional numbing, and difficulty returning to "normal" life after deployments. They describe the cumulative weight of witnessing—each disaster adds to the last. The faces blur but the feelings compound.[53] These are the same symptoms described by Graves Registration veterans across a century of conflict.

The parallel extends to moral injury. Red Cross workers, like military chaplains and medics, sometimes face situations in which they cannot help everyone, resources are insufficient, and the system fails the people it was designed to serve. A shelter that cannot accommodate everyone, a family that receives inadequate assistance, a death that could have been prevented—these are not combat wounds. They are moral wounds from knowing that something was wrong, and that one was powerless to fix it.[54]

I know this terrain from the inside. As an American Red Cross volunteer—an IHL educator, a Service to the Armed Forces volunteer, a participant in Restoring Family Links—I have stood at the edges of grief that mirrors what this chapter describes.[55] I have not served in Mortuary Affairs. I have not handled remains. But I have sat with families navigating military loss. I have taught the laws designed to protect the dead. I have helped connect people separated by conflict and disaster, and I have witnessed the particular anguish of not knowing.

That witnessing leaves marks.

Not the same marks as those who zipped the bags and dug the graves, but marks nonetheless. The Red Cross taught me that care is not free, that every act of witness costs something, and that the cost must be acknowledged or it compounds.

What military institutions are only now learning, disaster response organizations understood earlier: that caretakers need care. Debriefing is not weakness. Silence is not strength. The work of carrying others' grief requires support, language, and community.[56]

The moral injury framework that is finally being applied to Graves Registration veterans emerged, in part, from research on humanitarian workers, chaplains, and disaster responders.[57] The wounds are not identical, but they share a common root: the cost of caring in contexts in which care is desperately needed and inadequately supported.

This is not to equate Red Cross volunteerism with combat burial duty. The scales are different. The intensity is different. But the underlying truth is the same.

Those who carry others' grief must be carried themselves.

And too often, they are not.

Interlude: The Ones Who Stayed

"We didn't run toward glory. We stayed with the aftermath."—Vietn am-era Mortuary Affairs specialist

Some soldiers ran toward the bullets.Others ran toward the wo unded.But a few stayed behind where the shooting had stopped, and the silence had begun.

These were the ones who stayed. To zip the bags, to clean the blood, to tag the toes. To carry what was left.

They were not celebrated. They were not remembered. But they were necessary.

In that silence, in that aftermath, they became something more than witnesses—they became caretakers of grief.

And they carried that task long after the war went quiet.

Spotlight: Trench Ghosts and Night Terrors

The Sleepless Aftermath of Burial Work

They called them *trench ghosts*—the dreams that followed burial men home.They weren't dreams, really. They were Hauntings.

Veterans of the Graves Registration Service and Mortuary Affairs often spoke of nighttime visitations—sometimes in interviews, more often in whispers. They would wake sweating, shouting, certain that someone had touched them. Or that someone was missing. Or that they were back in the pit, with only a flashlight and a name to recover.

These were not metaphorical ghosts. They were *sensory*.

Men described the smell of decaying soil suddenly returning in their sleep. Others reported waking to phantom sounds: the snap of a body bag, the scrape of a shovel, the distant voice of a mother asking, "Was he alone?"

Psychologists now identify these experiences as night terrors associated with moral trauma.[58] But for the men who lived them, no label softened their impact. Some refused to sleep without light. Others avoided sleeping in silence. Many simply learned to live without sleep at all.

The condition was widespread. A World War II GRS veteran once said:

> *"I didn't sleep more than three hours at a time for thirty years. Every time I closed my eyes, I was back there. I could hear the mud sucking at my boots."*

Some described ghosts who never spoke. Others spoke too much.

A few dreamed in ledgers: long columns of names, with just one missing.

There were coping mechanisms—alcohol. Routine. Avoidance. But rarely relief. And almost never understanding. Families grew accustomed to closed bedroom doors and muffled cries. Spouses learned not to ask. Some left. Others stayed, becoming silent witnesses to the haunted.

These night terrors were not about guilt over survival. They were about being *too close* to death for too long, too often, and with too little space to breathe.

And because the world did not recognize these men as "combat-wounded," their sleeplessness was rarely treated as trauma. It was dismissed as stress or labeled as insomnia. Chalked up to age, or temperament, or bad dreams.

But they weren't dreams. They were the *echoes* of rituals performed too many times. They were the unfinished funerals of memory. They were the war, still asking to be buried.

Even now, veterans' homes across the country host former mortuary specialists who sleep lightly, if at all. Some still fold their clothes with military precision. Some still wake to the sound of voices no one else can hear.

They are not imagining the dead.They're remembering them.Be cause they were the last ones to know them.

And the body remembers what the world forgets.

Artifact: The Shrapnel Box

A Footlocker of Memory and Mourning

It looked like any other military footlocker. Standard-issue green.

Dented corners. Metal latch rusted from time. But inside, it was a war memorial in miniature.

The "Shrapnel Box," as one family came to call it, was found in the basement of a Korean War veteran's home after his death. He had served with Graves Registration, but he rarely spoke of it. The box spoke volumes.

Inside were artifacts not meant for public display:

- A laminated field map with numbered temporary gravesites
- Dozens of carbon-copied field forms, faded but still legible
- A rosary missing two beads
- A letter never mailed, its paper brittle, the envelope unsealed
- And in one folded shirt pocket: a child's marble, still faintly red

There were no medals, no photographs of the veteran himself. Only documents, fragments, and one short note taped to the lid:

"For the ones we could not save. For the ones we could not forget."

Boxes like this exist across basements and attics, folded into closets, wrapped in cloth. Some contain ledgers. Some contain relics. Some were burned in silence after a bad night. Others were tucked away and passed down to children who never fully understood.

In anthropology, such personal reliquaries are called mnemonic containers—objects that hold memory when words cannot.[59] In the context of war, they become sacred vessels, bridging the personal and the collective, the buried and the remembered.

To the military, these items may appear as detritus. To a museum, they may seem incomplete. But to the one who packed the box, they were a map of meaning.

They were grief, organized. Guilt, folded. Witness, made material. These boxes were not for public mourning. They were private sanctuaries of memory, shrines for the caretakers who had no one to mourn for *them*.

And in every clasp, every crease, every fingerprint still left on the lid is the simple plea these men carried home with them:

"Remember what we carried. Even when we couldn't speak it aloud."

The Weight of the Unnecessary

Not all deaths weigh the same.

This is a truth the caretakers know but rarely speak. The physical labor is identical—a body is a body, whether killed by artillery or execution, by disease or deliberate starvation. But the moral weight is different. And that difference compounds the trauma in ways that standard grief frameworks fail to address.

When a soldier dies in combat, the caretaker grieves. When a soldier dies because his general ordered an attack in the final minutes of a war already won, the caretaker grieves *and rages*. The death was unnecessary. The body did not have to be here. Someone made a choice, and that choice created this corpse.

On November 11, 1918, approximately 2,738 men died after the Armistice was signed—in the six hours between signature and impl ementation.[60] Commanders ordered attacks knowing the war would end at 11:00 a.m. Some sought last-minute glory. Some followed orders without question. Some simply couldn't imagine stopping. The result was the same: thousands of bodies that did not have to exist.

The Graves Registration soldiers who gathered those remains knew what the ledgers would later confirm. These men died for nothing. Not for victory—the victory was already secured. Not for defense—no position was at risk. They died because someone with authority chose to spend their lives in the war's final breath.

That knowledge does not leave a caretaker.

The same burden fell on the liberators of Nazi concentration camps. When British forces entered Bergen-Belsen in April 1945, they found a landscape of death beyond processing: 10,000 unburied bodies, 60,000 survivors ravaged by typhus and starvation, a camp so contaminated that traditional burial was impossible.[61] The decision was made to bulldoze bodies into mass graves and burn the barracks to prevent epidemic.

The soldiers who drove those bulldozers were not perpetrators. And they carried the weight of knowing that the dead they were bulldozing deserved better—but that the scale of atrocity had overwhelmed even death's most basic dignities.

This is the particular trauma of processing atrocity: the caretaker becomes witness not only to death but to *injustice*. They see what the perpetrators tried to hide. They restore identity to those who were meant to be anonymous. They carry the evidence that tribunals will later use.

The Nuremberg Trials established a principle that matters to every caretaker who has processed preventable death: individuals are accou ntable.[62]

"Following orders" is not a defense. The chain of responsibility extends from the hand that pulled the trigger to the signature that authorized the transport. When justice is rendered, it validates what the caretakers witnessed. It says: *you were right to be angry. This should not have happened. Someone is being held to account.*

But justice often fails. Perpetrators die in their beds. Tribunals are blocked by politics. History is rewritten by the victors. And the caretakers are left holding witness to atrocities that the world prefers to forget.

How do we care for those who carry this weight?

We name what they witnessed. We refuse to sanitize history into acceptable narratives. We hold perpetrators accountable whenever possible and acknowledge the failure when we cannot. We listen not just to their grief but to their outrage, understanding that anger at injustice is a healthy response to having processed its results.

And we continue their work. Every effort to prevent future atrocities—every IHL education program, every genocide early-warning system, every international tribunal—is an act of care for the caretakers who came before. We cannot undo what they witnessed. But we can work to ensure that fewer will inherit their burden.

The plague does not choose its victims. The earthquake does not hate. But the general who ordered the final charge, the bureaucrat who signed the transport manifest, the government that turned away the refugee ship—they chose.

And someone had to bury what they made.

The caretakers remember. We must remember them.

THE LIVING DEAD

"I didn't die there. But part of me did."—GRS veteran, interview, 2002

They survived.But they were never the same.

The men of the Graves Registration Service and Mortuary Affairs walked among the dead so others didn't have to. They performed the last kindness a soldier can receive—the dignity of being named, recorded, and returned. They buried the war, one body at a time. But in doing so, they buried something inside themselves.

And when they came home, they carried the silence of that burial with them.

They were haunted not by ghosts in the traditional sense, but by *records*. By names written over and over, by smells that lived in their sinuses, by questions that were never answered—"Did he suffer?" "Was he alone?"

They were the last to touch the dead and the first to lose language for their own pain.

And yet.

There is a man in his eighties who still wakes at 3:00 a.m., certain he has forgotten a name. He reaches for a notebook that isn't there anymore—the one he carried in 1944, the one his wife threw away after he woke screaming too many times. He lies in the dark and recites names anyway, whispering them like prayers, because stopping feels like abandonment.[63]

There is a woman in her sixties who never knew why her father couldn't attend her graduation, her wedding, her children's baptisms—any ceremony with flowers and solemnity. She found out after he died, when she opened the footlocker. She is still finding out. She will be finding out for the rest of her life.[64]

There is a VA counselor who specializes in moral injury who has spent twenty years trying to convince the institution that burial workers deserve the same recognition as combat veterans. She has succeeded, partially. The diagnostic criteria are expanding. The research is growing. But she knows that for every veteran who finally receives care, there are hundreds who died without it—and thousands who never knew care was possible.[65]

There is a question that has no answer:

What do we owe those who carried the dead?

We owe them what they gave the dead: recognition, remembrance, dignity.

But we cannot give them rest. The weight does not lift. The names do not fade. The silence does not break simply because someone finally asks.

A system is beginning to listen—that is true. Language is emerging, and chaplains are trained differently. Mortuary Affairs units receive psychological support that did not exist a generation ago.[66] These are gains, hard-won and incomplete.

But most of the men who carried the dead in 1944, in 1951, and 1968 are gone. They died carrying what they could not speak. And the few who remain are very old now, sitting in quiet rooms, remembering names that no one else does.

This chapter has tried to name what they carried. It has not tried to heal it—healing is not this book's to give. What this book can offer is witness. The same witness they offered the dead.

To say: *you were here. You did this. It mattered. It cost you.*

And: *we see you now. Even if we saw you too late.*

The burden of memory does not end with acknowledgment.

But acknowledgment is where it begins.

Interlude: What They Made

Some deaths are given. Some deaths are made.

The plague does not choose. The earthquake does not hate.

But the general who ordered the charge at 10:45 a.m. on November 11, 1918—knowing the Armistice would take effect at eleven—he chose.

The bureaucrat who signed the transport manifest to Auschwitz, who scheduled the trains, who calculated the capacity of the chambers—he chose.

The government that turned away the refugee ship, that closed the borders, that looked away while the smoke rose—they chose.

And when the choosing was done, someone had to gather the bodies.

The caretakers did not choose.

They did not choose the last-minute assault, the mass grave, the crematorium backlog. They did not choose the impossible scale, the contaminated ground, the faces frozen mid-scream. They did not choose to inherit what cruelty had made.

But they showed up with shovels and stretchers and silence. They showed up because the dead were waiting. They showed up because no one else would.

And they buried what did not have to exist: the unnecessary dead. The preventable dead. The dead who were someone's decision. They buried them alongside the inevitable—the ones who had fallen in honest battle, the ones whom medicine could not save, the ones whose time had simply come.

They buried them all the same. But they knew the difference.

A death that had to happen weighs like grief. A death that did not have to happen weighs like injustice.

The caretakers carry both.

And the injustice does not lift. Not with time, not with ceremony, not with the verdict at Nuremberg. It lifts only when someone says: *this should not have happened. Someone is responsible. And we remem-*

ber.

The caretakers remember. They remember the time stamps: 10:58 a.m. They remember the bulldozer blade. They remember the transport manifest. They remember what the perpetrators tried to bury.

And now we remember them. The ones who buried what was made. The ones who carried injustice in their hands. The ones who knew.

They knew. And they stayed anyway.

Reflection

The cemetery was closing, but they let us stay.

We had come to Belgium following the trail of World War I—visiting Ypres, walking the trenches, standing in the museum where the horrors of industrial warfare are preserved behind glass. On our way back to the hotel, we saw a sign: "Flanders Field American Cemetery."

I knew I needed to stop.

When we arrived, I couldn't speak. The rows of white crosses stretched endlessly across manicured grass—young men from a war that had ended over a century ago, still tended, still honored, still remembered. My eyes welled up. My throat closed. I walked among the graves in silence, reading names I did not know but somehow felt I owed.

The Belgians, I learned, adopt these graves. Families take responsibility for individual soldiers, laying flowers, maintaining the site, passing the commitment to their children and grandchildren. They celebrate American Memorial Day. They have not forgotten what these young men did for their country.

They remember when we do not.

As the cemetery prepared to close, the staff noticed we were still there. Rather than asking us to leave, they played "Taps," the lone trumpet echoing across the stones as the sun began to set. We found headstones of young soldiers from Utah, our home state, who had died just days before the Armistice. Boys who had almost made it home.

I fell to my knees.

The care of our dead in another country, by people who have no obligation except the one they have chosen—it broke something open in me. This is what memory looks like when it is tended. This is what honor looks like when it is practiced, not just proclaimed.

I left Flanders Field changed. The book I would eventually write began to take shape that afternoon—not as an academic exercise, but as an act of witness. A response to a question I could no longer ignore:

If the Belgians can remember our dead, why can't we remember the ones who buried them?

CHAPTER EIGHT

ANTHROPOLOGY OF WAR DEATH

"Before the flags, before the politics, there was ritual. We buried our dead because we were human."—Military anthropologist, field notes, 1999.

Long before there were uniforms or casualty reports, long before there were flags to salute or dog tags to recover, there were bod ies.And there were those who cared for them.

This chapter turns back the clock. Not to erase military history, but to uncover what underlies it: the deep, shared instincts that have always shaped how human beings respond to war death. From the burned offerings of Bronze-Age battlefields to scaffold burials in Indigenous North America, from relics gathered off enemy warriors to sacred songs sung over unnamed remains, the dead have always asked something of the living.

We will explore what it means to bury the war dead across cultures—and what that burial says about who we are. We examine ritual as identity, forensic science as witness, and the shared moral landscapes that transcend flags and factions. We will look at Indigenous perspectives on warrior death, the evolution of naming the unknown, and moments where enemies found common humanity in the act of burial.

This is not a chapter about how wars are fought.It is about how they end.And how they are remembered.

Because in every war—regardless of geography or century—there is one act that returns us to ourselves: we bury the dead.

Not just for their sake, but for ours.

Interlude: Ash and Offering

"Even in war, we stopped to bury. We always stopped to bury."—Bron ze-Age battlefield excavation notes, 2012

They found him curled beside a shield. No armor, no boots, just bone. But around him—carefully arranged—were remnants of meaning: a broken blade, a charred seed pod, and what appeared to be ochre smeared across his skull.

He had not been left behind. He had been *buried*.

Three thousand years ago, in a time before maps or manifestos, someone knelt over this man and made a choice.To honor him.To send him home.To turn death into memory.

Archaeologists call it ritual burial. But in the language of the heart, it is something simpler: refusal. Refusal to let death go unmarked, to let violence erase identity, to become animals, even in the midst of carnage.

They found similar signs on other bodies: carved beads near the throat, food offerings at the feet, ashes smudged on foreheads. A battlefield, yes, but also a cemetery. A place of death, yet a place of reverence.

There is no surviving scroll to tell us why they did it, but their actions speak.

"This one mattered. We still remember how to care."

To Bury Is to Be Human: Ritual as Identity

"When we take a life in battle, we return it to the earth with honor. That is our way."—Pacific Islander warrior tradition

Long before there were nations—before borders, medals, or marble tombs—there were bodies. And there were those who cared for them.

Across continents and centuries, humans have responded to death not with indifference but with ritual. Burial is not merely a tradition; it is a declaration. It marks the line between life and memory, between flesh and meaning. It declares that a person mattered. And we still remember how to care.

Anthropologists widely consider ritual burial one of the earliest signs of sentient human behavior.[1] In sites such as Shanidar Cave in Iraq, the remains of Neanderthals dating back over 60,000 years have been found interred with pollen traces, suggesting intentional floral offerings.[2] These early acts of mourning speak to the same impulse that drives modern military funerals: to provide dignity, to mark a passage, to create continuity between the living and the dead.

In wartime, this impulse becomes a form of resistance.

Combat strips away humanity. The noise, the brutality, and the dehumanization of the enemy work to sever identity from the body. And yet, even amid this chaos, people return to the bodies. They dig. They cover. They speak. They light fires. They whisper names.

To bury in war is to insist that violence will not write the final chapter. It is a refusal to let the battlefield become the only memory of the fallen.

From Iron-Age warrior burials in Scandinavia, with horses and weapons placed beside the dead, to scaffold burials among Plains Indigenous nations, to modern-day repatriation ceremonies for unidentified soldiers, the act of caring for the war dead serves both spiritual and social roles.[3] It reaffirms the values of the living while acknowledging the cost of conflict.

These rituals transcend ideology. They are found on all sides of war.

In ancient Greece, truces were often arranged after battles so both sides could retrieve their dead.[4] Among the Māori of Aotearoa (New Zealand), warriors were buried with extensive ceremony, regardless of their side, to ensure spiritual balance was maintained.[5] In Islam, even in battle, the bodies of enemies must be buried with respect. And across military traditions, chaplains of opposing faiths have prayed over enemy graves in recognition of a shared humanity.

Burial rituals are not just about the dead.They are about the living. They are how societies declare who they are and who they refuse to become.

Even in modern war zones, soldiers still pause. Medics still cover the fallen. Volunteers dig in silence. A boot is laid upright, a name is written. Small but profound gestures that say: we may be surrounded by death, but we have not forgotten how to be human.

Because to bury the dead is not just an end.

It is a beginning.

A beginning of memory.A thread back to meaning.A way to hold the soul, even after the body is gone.

From Dog Tags to DNA: The Evolution of War Identification

"We used to carry their names around their necks. Now we carry them in our code."—Mortuary Affairs officer, 2009

In the wake of war, the first question is often the most painful: who lies here?

Throughout history, soldiers have died unnamed and unclaimed—buried hastily in trenches, left on the battlefield, or recorded only in memory. But as war became industrialized and death became bureaucratized, the demand to identify the fallen became a sacred priority. Identification is not merely administrative. It is a final act of care, a return of identity to the dead.

Dog Tags and Disks: The Early Tools of Identity

During the U.S. Civil War, soldiers often resorted to writing their names on scraps of paper or carving initials into their belts, knowing that the odds of dying unidentified were high. It was not until 1906 that the U.S. Army issued official identification tags—circular aluminum disks stamped with a soldier's name and unit.[6]

By World War I, soldiers wore two tags: one to stay with the body, one to return with the casualty report. But conditions in the trenches often made recovery impossible. Bodies decayed beyond recognition. Disks were lost. And the war's scale overwhelmed early identification systems. In the postwar period, vast cemeteries of unnamed soldiers were created, many marked only as "Known Unto God."

World War II standardized the modern dog tag. They were rectangular, made from steel, and etched with vital information: name, blood type, religion, and service number.[7] They were worn on ball chains around the neck. Though more durable, they could still be destroyed by fire or buried beyond recovery. When dog tags failed, medics and mortuary teams relied on dental records, uniform fragments, personal effects, even handwriting in notebooks to reconstruct identities.

The Silent Science: Fingerprints, Forensics, and the Paper Trail

By the Korean War, the U.S. military had implemented fingerprinting and retained dental charts for every service member.[8] These became critical in cases where remains were damaged or decomposed. Field teams carried fingerprint kits and photographic tools, and mortuary personnel received training in cataloging skeletal remains.

During the Vietnam War, autopsies became more common, especially for unidentified or partial remains. Mortuary specialists used anatomical diagrams to document injuries and confirm identities. Despite these advances, more than 1,500 U.S. service members from Vietnam remain unaccounted for.[9]

Every tag, fingerprint, or personal belonging was a chance to restore not just a name but a story.

The DPAA: Burial Work Extended Across Generations

"We don't recover remains. We recover people." —DPAA forensic anthropologist, 2019

The work that began with shovels in Civil War fields continued with ledgers in World War I trenches and expanded with dog tags across World War II battlefields, though it did not end when those wars concluded. It continues today in laboratories, on archaeological digs, and in the patient, painstaking process of returning names to the missing.

The Defense POW/MIA Accounting Agency (DPAA) represents the institutional embodiment of the promise that no American service member will be forgotten, no matter how long they have been lost.

Established in 2015 through the merger of several predecessor organizations, the DPAA is the world's largest forensic identification operation dedicated to military personnel.[10] Its mission spans eight decades of conflict and reaches across the globe—from Pacific islands where World War II aircraft went down, to Korean hillsides where soldiers fell in 1950, to Vietnamese jungles where remains still emerge from the earth.

More than 81,500 Americans remain missing from past wars and conflicts.[11]

The numbers are staggering:

- Over 72,000 from World War II
- More than 7,500 from the Korean War
- Approximately 1,580 from the Vietnam War
- Smaller numbers from the Cold War and more recent conflicts

For each of these missing, somewhere there is—or was—a family that never received a body, never held a funeral, never had a grave to visit.

The DPAA's work combines multiple disciplines into a single mission. Historians research archival records, eyewitness accounts, and declassified documents to locate probable burial sites. Archaeologists conduct excavations using techniques developed for ancient sites, applied to 20th-century battlefields. Forensic anthropologists analyze skeletal remains, reading bones for age, ancestry, stature, and trauma. Odontologists compare dental records. And DNA specialists extract genetic material from degraded remains, matching it against reference samples provided by families, some of whom have waited seventy years for this moment.[12]

The field teams are, in essence, burial crews working in reverse. Where GRS personnel once gathered the dead and placed them in the ground, DPAA teams exhume remains and restore them to identity. The work requires the same qualities: physical endurance, emotional resilience, meticulous attention to detail, and profound respect for the dead.

A typical recovery mission might involve weeks of preparation, followed by months of excavation at a remote site—a jungle crash location in Laos, a Pacific island battlefield, a former POW camp in North Korea.[13] Teams sift through soil with archaeological precision, documenting every fragment, every artifact, every possible clue. The remains they recover are transported to the DPAA laboratory in Hawaii, where the slow process of identification begins.

Some identifications come quickly. DNA matches are found, dental records align perfectly, personal effects confirm beyond doubt. Others take years. Some may never be completed. But the agency's commitment remains: as long as recovery is possible, the work continues.

The families who finally receive word, sometimes decades after they stopped expecting it, describe the experience as both ending and beginning. The long uncertainty is over. The questions are answered. But now comes the grief that was suspended, the funeral that was postponed, the burial that can finally occur.[14]

One mother, notified in 2018 that her brother's remains had been identified from a Korean War battlefield, said: "We always knew he was out there. We just didn't know where. Now we can bring him home and tell him we never forgot."[15]

The DPAA represents something profound about American military culture: the belief that the obligation to the dead does not expire. That a promise made in 1944 or 1951 or 1968 remains binding in 2024. That burial work is not complete until every possible recovery has been attempted, every possible identification pursued.

This is not logistics. This is moral commitment extended across generations.

The scientists and archaeologists and analysts who perform this work are caretakers, no different, in essence, from the GRS soldiers who dug graves in France or processed remains in Vietnam. They handle the dead with reverence. They speak the names. They return what was lost.

They are the final link in a chain that stretches back to the first military burial and forward to the last family waiting for word.

The dead are patient. The DPAA is faithful. And the work continues.

DNA and the Rise of the Modern Unknown

The 1990s ushered in a quiet revolution in identification: DNA forensics.

Following advances in molecular biology, the Department of Defense established the Armed Forces DNA Identification Laboratory (AFDIL), which began building a database of reference samples from active-duty personnel and family members of the missing.[16] DNA became a last resort for identifying the most fragmented or degraded remains when all else had failed.

A watershed moment came in 1998, when the remains from the Vietnam Unknown in the Tomb of the Unknown Soldier were exhumed and identified through mitochondrial DNA as 1st Lt. Michael Blassie, an Air Force pilot shot down in 1972.[17] His remains were returned to his family, and the tomb was left symbolically empty—its purpose fulfilled, and yet eternal.

Today, DNA identification is standard in Mortuary Affairs protocols. Forensic teams in Iraq and Afghanistan used DNA to confirm remains even from IED blasts or fire-damaged sites. The science has also been retroactively applied to World War II remains discovered in Europe or the Pacific. Families who waited for decades have finally received confirmation, and with it—closure.

The Soul of Identification: Why It Matters

Identification is not just a technical process. It is a sacred promise.

Every dog tag pressed, every tooth chart filed, every swab of bone dust in a lab—these are acts of reverence. They say to the dead: you will not vanish. You will be named.

For the living, identification restores not only a body, but a bond. It allows a mother to bury her son, a child to grieve a father, a community to raise a flag with more than a silhouette.

In every era, from makeshift battlefield graves to high-tech forensic labs, this truth remains:We fight for names because they carry meaning. And those who identify the dead are not merely clerks or scientists. They are custodians of identity, and they give voice to the fallen. They refuse to let memory dissolve into dust.

The Enemy Dead: Cross Cultural Practices in Combat

"The war ends when the enemy is buried."—Zulu proverb

War divides us by blood, uniform, and flag. But death—death has a strange way of making us equal again.

Across centuries and civilizations, warriors have laid down their arms, if only briefly, to retrieve the fallen. Not only their own, but their enemies. Even amid conflict, the urge to bury persists. And in these moments, ritual becomes a bridge: not of peace, but of recognition.

Ancient Respect for the Fallen Foe

In Homer's *Iliad*, the vengeful Achilles, having slain Hector, is moved to return his body to King Priam.[18] It is not mercy that compels him, but the gravity of death itself. In classical Greece, truce days were often declared after battle so both sides could recover the dead.

Denying burial was considered not only dishonorable but sacrilegious—an affront to both the gods and to human order.[19]

The ancient Persians similarly honored enemy dead, especially noble warriors. Xenophon recounts how Persian commanders allowed Greek troops to bury their fallen after battle, even assisting with the rites.[20] Such moments transcended politics and strategy—they acknowledged a shared sacred obligation.

Among the Mongols, who practiced sky burials and guarded warrior graves with secrecy, worthy foes were sometimes buried with honors, their weapons treated as relics.[21]

Indigenous Reverence for All Dead

Across Indigenous cultures of the Americas, death in war was not solely a military outcome—it was a spiritual crossing. For the Lakota, even dead enemies were sometimes cleansed, sung over, and returned to their people with offerings or items of meaning.[22] The Cherokee believed that to mishandle the enemy dead could disrupt the balance between the physical and spiritual world, leading to disharmony and misfortune.[23]

The Haudenosaunee (Iroquois Confederacy) viewed mourning and condolence rituals as a sacred responsibility, extending them even to those who died by their hand. These were not acts of concession; they were expressions of cosmology and ethical strength.

In the Hawaiian warrior tradition, even fallen enemies carried *mana*—spiritual power. To desecrate their remains risked unleashing spiritual imbalance. Chiefs were often buried in secret to protect this sacred force, and enemy warriors were sometimes given burial to prevent chaos from reverberating into the land.[24]

Mutual Humanity in Modern Wars

During World War II, Allied chaplains in Europe were sometimes recorded praying over both Allied and Axis bodies. In the Pacific, Japanese soldiers frequently carried deeply personal items—family crests, ofuda prayer slips, or locks of their children's hair. Some American soldiers, moved by instinct rather than order, buried these items with the bodies instead of collecting them as trophies.[25]

Yet contradictions abounded. In both the European and Pacific Theaters, enemy corpses were desecrated, photographed, or used for propaganda. These acts were justified as dehumanizing the other, but they often left a spiritual cost in their wake, especially for those tasked with burial duties.

In Korea and Vietnam, ceasefires were occasionally observed to retrieve and exchange the dead. Many soldiers recall quietly assisting locals or even enemies in retrieving fallen comrades—without words, just gestures.[26]

Today, humanitarian organizations such as the International Committee of the Red Cross (ICRC) and local NGOs negotiate body exchanges in active conflicts—in Syria, Ukraine, Israel-Gaza, and more. These acts, while fraught and fragile, remind us of a buried truth: to bury the enemy dead is not betrayal. It is testimony.

The Moral Act of Burial

Across anthropology and faith traditions, burial is more than disposal. It is spiritual diplomacy, a kind of absolution. Even when nations refuse to speak, the earth still accepts the body.

Some soldiers report having felt changed after burying an enemy. The helmet still wore the wrong emblem. The hands once held a rifle. But in death, those distinctions blurred. All that remained was a human form, still and fragile—and a silence too heavy to ignore.

"He was my enemy," one soldier wrote. "But once he died, he was only a man."[27]

Bones Tell Stories: Forensics, Relics, and the Truth of the Body

"The dead do not speak. But their bones do."—Forensic anthropologist, exhumation site, Bosnia

When all else is lost—uniform, dog tags, memory—bones remain.

In war, where chaos can obliterate identity, bones become the final archive. They hold trauma like text, carry culture in form, and whisper names long thought silenced. They are not just biological remains; they are narrative, grief, and evidence fused into matter.

The Rise of Forensic Anthropology in War Zones

Forensic anthropology—the study of human remains to determine identity and cause of death—became essential during and after the 20th century's most brutal conflicts. Its roots trace to archaeologists and anatomists assisting with mass grave recovery in postwar Europe and Latin America. Today, it is a multidisciplinary field combining osteology, pathology, DNA analysis, and cultural sensitivity.

In places like Rwanda, Bosnia, and Cambodia, forensic teams work under international tribunals to uncover the truth of genocides and war crimes.[28] In Vietnam, Southeast Asia, and the Pacific, U.S. recovery missions are ongoing, with anthropologists operating field labs to examine bone fragments, teeth, and relics.[29] Every mark, fracture, or healing pattern holds clues—trauma, age, handedness, disease, and sometimes even occupation.

Bones tell us who someone was. But they also tell us how they died.

A clean cut may indicate execution. A blast radius tells of IEDs or mortars. Fractured wrists suggest restraint. Dental wear may signal poverty, malnutrition, or cultural practices. These clues help build a profile, moving one step closer to restoring a name.

The Red Cross and the Work of Witnessing

The International Committee of the Red Cross (ICRC) has long recognized that identification of the dead is not merely a forensic matter but a humanitarian imperative. Since the aftermath of World War I, when the ICRC helped establish prisoner-of-war information bureaus and tracing services, the organization has understood that families separated by conflict need answers, and that the dead deserve names.[30]

Today, the ICRC's forensic programs operate in conflict zones around the world, assisting with the recovery and identification of the dead from Syria to Colombia to the former Yugoslavia.[31] The organization trains local authorities in proper handling of remains, advocates for the rights of families to know the fate of missing relatives, and documents the dead with the same meticulous care that characterized Clara Barton's Office of Missing Soldiers more than 150 years ago.

The work intersects with forensic anthropology at every level. ICRC protocols inform how mass graves are excavated. ICRC documentation standards shape how remains are catalogued. And the ICRC's insistence on the dignity of the dead, enshrined in International Humanitarian Law, provides the ethical framework within which forensic science operates.[32]

Identification is not a favor granted to families. It is a right protected by law and a duty borne by those who claim to uphold humanitarian values.

The forensic anthropologist brushing soil from a bone, the DPAA specialist analyzing DNA, the ICRC delegate documenting a gravesite—all are ensuring that the dead are not forgotten, that families receive answers, and that the chaos of war does not erase the individuals it claims.

Relics of the Fallen: Personal Objects and Sacred Symbols

Sometimes, bones are not found alone. They lie beside talismans of life:

– A wedding band– A crumpled photograph– A necklace woven by a child– A prayer slip or tribal charm

These items may offer more than forensic help. They offer emotion. Identity is not just biological—it is relational. To return a ring to a widow, or a necklace to a child, is to return a part of someone's story.

Yet the handling of such relics raises ethical dilemmas. Should they be preserved for evidence? Given to families? Displayed in museums? Or buried again, as sacred remains?

In many Indigenous traditions, bones and belongings are considered spirit-bound. Their disturbance without ceremony is taboo. The Native American Graves Protection and Repatriation Act (NAGPRA), passed in 1990, mandated the return of ancestral remains and sacred items from U.S. institutions to their tribal nations.[33] Similar repatriation efforts have emerged worldwide, fueled by activism, archaeology, and a reckoning with colonial violence.

When Cultures Collide: The Ethics of Disturbing the Dead

"You call it science. We call it desecration."—Indigenous elder, NAGPRA consultation, 1995

The same forensic techniques that restore names to missing soldiers have, in other contexts, caused profound harm. The history of anthropology is entangled with colonialism, and the study of human remains is no exception. Before burial work could become healing work, it first had to reckon with its own violations.

For more than a century, Western scientists collected Indigenous remains from battlefields, from burial grounds, from sacred sites and stored them in museums, universities, and private collections.[34] These were not anonymous bones. They were ancestors, removed without consent, displayed without dignity, and studied without regard for the spiritual beliefs of their descendants.

The justifications were familiar: scientific progress, the preservation of "vanishing" cultures, the advancement of knowledge. But the effects were devastating. Communities lost access to their ancestors. Spiritual practices were disrupted. The dead, who should have been at rest, were trapped in drawers and display cases.[35]

The Native American Graves Protection and Repatriation Act (NAGPRA), passed in 1990, became a turning point.[36] The law required federally funded institutions to inventory their holdings of Indigenous remains and sacred objects, consult with descendant communities, and, upon request, return what had been taken.

The process has been slow, contested, and incomplete. Some institutions resisted; some remains lacked clear provenance, making repatriation legally complex. Some tribes lacked the resources to pursue claims. But tens of thousands of ancestors have been returned, and with them, the possibility of proper burial.[37]

For forensic anthropologists working on military recovery, NAGPRA established principles that extend beyond Indigenous remains. It affirmed that the dead have rights—or, more precisely, that the living have obligations to the dead that science cannot override. It established that cultural and spiritual beliefs matter in determining how remains are treated. And it demonstrated that "identification" is not merely a technical process but an ethical one.

The tensions remain. When remains are discovered—whether Indigenous ancestors or military missing—decisions must be made about whether to excavate, how to handle what is found, and who has authority over the process. These decisions are not purely scientific but are moral, spiritual, and political.

Some Indigenous communities have declined forensic analysis of recovered remains, preferring to rebury without identification rather than subject ancestors to further disturbance.[38] This choice can be difficult for families and agencies accustomed to the Western imperative of naming. But it reflects a different understanding: that the dead deserve rest more than they deserve labels, that spiritual wholeness matters more than documentary certainty.

Other communities have embraced forensic collaboration, seeing identification as a form of restoration—returning not just bones but stories, connecting descendants to ancestors whose names had been lost.[39] The key, in both cases, is allowing affected communities to guide the process rather than having decisions imposed upon them.

This is what listening as care looks like.

It means recognizing that Western forensic practice is not neutral—that it carries assumptions about the body, the soul, and the meaning of death that are not universal. It means approaching every recovery with humility, acknowledging that the dead belong not to science but to those who loved them, and to the traditions that give their deaths meaning.

The work of identification is sacred. But it is sacred only when performed with consent, with respect, and with awareness that bones are not just evidence, they are people.

And people deserve to be asked how they want to be remembered.

When Burial is Denied: The Armenian Genocide

Sometimes the atrocity includes the denial of burial itself.

During the Armenian Genocide of 1915–1923, the Ottoman government did not merely kill, it erased. An estimated 1.5 million Armenians perished through massacre, forced death marches into the Syrian desert, starvation, and exposure.[40] But the genocide extended beyond the killing. Bodies were left unburied in the desert, thrown into the Euphrates River, dumped into caves and wells. The denial of proper burial was deliberate, a final act of erasure designed to leave no graves, no markers, no evidence that a people had existed.[41]

For Armenian families, this meant a grief without location. There were no cemeteries to visit, no headstones to tend, no sacred ground where memory could be anchored. The dead had been made to disappear not once but twice: first in body, then in burial.

The caretakers of the Armenian dead were, in many cases, the survivors themselves—those who had witnessed the marches, who had tried to bury family members with bare hands in the desert, who had marked locations in their memories that they would never be able to return to. Some survivors carried small objects—a photograph, a piece of jewelry, a fragment of clothing. These became the only graves they had.[42]

A century later, descendants are still seeking acknowledgment. Not just political recognition of the genocide (though that struggle continues) but the simple right to mourn. To know where. To have a place. The absence of burial created an inheritance of unfinished grief that has passed through generations.

In recent decades, some Armenian communities have conducted symbolic reburials. In these ceremonies, the dead are honored without bodies, ground is consecrated where no remains lie, and rituals are performed that their ancestors were denied.[43] These are acts of spiritual repair, attempts to complete what genocide left unfinished.

The Armenian case illustrates a terrible truth: burial can be weaponized. The refusal to allow care for the dead is itself a form of violence that extends the atrocity across generations. When caretakers are prevented from doing their work, the wound does not close. It passes forward, waiting for someone to finally say the prayers, to finally mark the ground, to finally let the dead rest.

The Armenians are still waiting.

And in that waiting, they join the families of the disappeared from Argentina, the missing from Syria, the unidentified from mass graves across the world. They are bound by a common inheritance: the knowledge that somewhere, their dead lie uncared for. And the determination that this will not be the final word.

Artifact: The Gweagal Shield – A Wound in Wood

It is a piece of wood, roughly oval, about three feet long. Its surface is weathered, scarred by age and salt air. At its center is a hole—clean-edged, unmistakable.

A bullet hole.

The Gweagal Shield has resided in the British Museum since the late 18th century, one of thousands of objects collected during Captain James Cook's voyages of Pacific exploration.[44] It was taken on April 29, 1770, at Kamay (Botany Bay), during the first recorded violent encounter between British sailors and the Aboriginal peoples of Australia.

According to Cook's own journal, two Gweagal warriors stood on the shore as the British boats approached. They held spears and shields, and they did not retreat. When the British fired, first with small shot, then with ball, one warrior was wounded. The other dropped his shield and ran.[45] The British collected what had been left behind: spears, a shield, and the opening chapter of two centuries of dispossession.

To the British Museum, the shield is an artifact. It is a specimen of Aboriginal craftsmanship, evidence of early contact, a curiosity from the age of exploration.

To the Gweagal people, it is something else entirely.

The shield was not merely a weapon. In Aboriginal culture, shields carried spiritual significance, often decorated with designs that connected the bearer to country, ancestors, and dreaming.[46] To carry a shield was to carry protection not just from physical harm but from spiritual imbalance. The shield mediated between the warrior and the forces arrayed against him.

The bullet hole at its center is not simply damage. It is a wound in wood, in history, in the relationship between peoples. It marks the moment when Aboriginal sovereignty was first physically contested by European force. It is, in a sense, the first casualty of a war that would last generations.

For decades, Gweagal descendants and Aboriginal activists have called for the shield's return.[47] The British Museum, citing its policy against repatriation, has resisted, though it has engaged in consultations and agreed to a long-term loan of the shield to Australia for exhibition in 2015.[48] The negotiations continue.

The case of the Gweagal Shield illuminates the broader tensions this chapter has explored. Who owns the dead—and the objects of the dead? What obligations do institutions bear to the descendants of those from whom they collected? When does preservation become possession, and when does scholarship become theft?

The shield asks these questions silently, from behind museum glass.

It cannot speak. But the hole at its center says enough:

Something was taken here.

Something was broken.

Something waits to be made whole.

Like the bones of soldiers buried far from home, like the sacred objects removed from graves and displayed in cases, the Gweagal Shield represents what happens when care is denied, when the dead and their belongings are treated as specimens rather than as sacred.

It is not just an artifact of first contact. It is a memorial to a moment when worlds collided, and when one world claimed the right to take what belonged to another.

The Gweagal people are still waiting for its return.

And the shield, with its wound still visible, is still waiting too.

The Bones that Refuse to Rest

In Germany and Eastern Europe, mass graves from World War II are still being found. In Vietnam and Korea, hundreds of thousands remain missing. Sometimes, they are discovered by farmers, construction workers, or during flood recovery.[49] In Guatemala and Argentina, forensic teams have unearthed sites tied to disappearances under dictatorships.

Each bone recovered is an act of resistance—against forgetting, against denial.

In this context, forensic anthropology becomes not just a science but a moral act. It declares that a life mattered. That a death has a truth. And that a body deserves more than silence.

Even the bones of the unknown carry story. They remind us that history has weight, that memory is material, and that the dead do not belong to politics—but to the living, who must choose how to remember.

"Every time I lift a femur from the dirt," one field technician said, "I am lifting a prayer someone else forgot to say."[50]

Known Unto God – The Tomb of the Unknown Soldier

"Here rests in honored glory an American soldier known but to God."
—Inscription, Tomb of the Unknown Soldier

In a nation that names its heroes, this one lies in silence.

The Tomb of the Unknown Soldier stands as both monument and mystery. It is a grave for the nameless, a shrine for the forgotten, and a mirror for every family who never received a letter, a body, or closure. In its marble stillness, it holds a truth: not all the dead are remembered by name. But all deserve remembrance.

A Monument to Absence

The idea of burying an unknown warrior is not unique to the United States. It was born out of the aftermath of World War I, a war so massive in its scale and so violent in its nature that thousands of soldiers were never identified—blown apart, buried hastily, or lost in mud and gas. France, Britain, and Italy each interred one such soldier in a national monument, meant to represent all the unidentified dead.[51]

The United States followed suit in 1921. From four exhumed bodies recovered in France, one was chosen at random by Sgt. Edward F. Younger, a decorated combat veteran. He laid a spray of white roses on the casket that "felt right."[52] That casket became America's Unknown Soldier, transported across the Atlantic aboard the USS *Olympia*, and buried in Arlington National Cemetery with full honors.

A president knelt.A nation wept.And a sentinel was posted to a watch that has never ended.

The Meaning Behind the March

Since 1937, soldiers of the 3rd U.S. Infantry Regiment—the "Old Guard"—have stood watch over the Tomb twenty-four hours a day, 365 days a year, through hurricanes, snow, fire, and national crisis. The ritual is precise: twenty-one steps, twenty-one seconds of pause, repeated with unwavering reverence.

It is the military's highest form of respect, the echo of a twenty-one-gun salute, performed not for an audience, but for a promise. These guards do not stand for applause. They stand because someone must.

They do not know the name of the man beneath their boots, but they honor him all the same.

And with him, they honor every soldier buried without a name. Every mother who never knew where her son fell. Every family whose grief was marked not by funeral but by silence.

When Names Are Found

In 1998, the Vietnam Unknown—interred alongside his World War I, II, and Korean counterparts—was disinterred and identified using DNA analysis as First Lieutenant Michael Blassie, an Air Force pilot shot down in 1972.[53] His family finally brought him home.

The empty tomb left behind now honors all unidentified soldiers of the Vietnam War. It is both sacred and unresolved. And it asks us to consider was it means to honor the unknown in an age when technology can recover almost every trace.

Perhaps the answer lies not in the science but in the symbol.

The Unknown stands for the unreachable. The unfindable. The unnamed losses that stretch across every war. Even today, thousands remain missing from past conflicts—buried in foreign soil, lost at sea, or scattered across unmarked graves.[54]

The Quiet Hands Who Remembered

What we often forget is this: before the Unknown could be laid to rest, someone had to find him. Someone had to lift the body from a trench, wash the mud from his boots, and prepare the coffin. Someone had to carry that coffin with care, knowing he had no name but deserved every dignity.

The tomb does not only memorialize the fallen. It memorializes the caretakers—the gravediggers, the morticians, the clerks, the chaplains—who ensured even the unnamed were not abandoned.

These quiet hands gave war its final grace. And in return, the Tomb offers them a silent honor. No spotlight, no citation. Only the certainty that someone noticed. Someone stood guard.

"We do not know who lies here," the inscription reads, "but we will not forget him."

The Tomb is not empty. It is full of memory and full of meaning. Full of the weight of all that war has taken and all that caretakers have tried to give back.

Ancestors and Afterlife: Indigenous Perspectives on War Death

"When a warrior dies, he returns to the people—not only in memory, but in spirit."—Lakota oral tradition

In many Indigenous cultures, the dead do not depart.They transform.

They become rain, mountains, bone, and wind. They are sung to, danced with, fed, and consulted. They live within the breath of the people, not as relics of the past, but as present ancestors, intimately woven into the community's continuity and health. Nowhere is this truer than in the realm of war death, where fallen warriors are not merely mourned, but honored as sacred.

A Death That Is Not an Ending

In contrast to many Western traditions, which often treat death as a final severance, Indigenous cosmologies often frame it as a transition—a journey from one realm to another. Burial is not the end of a life, but the continuation of that life's participation in the spiritual ecosystem.

Among the Lakota, warriors were traditionally scaffold-buried—elevated on wooden platforms facing the sky, wrapped in buffalo hides, and adorned with sacred colors. This placement allowed their spirit to ascend freely, remaining close enough to the people to offer protection and wisdom.[55] Songs were sung to recount their deeds. Families tended the scaffold with food and offerings.

In Andean traditions, such as those of the Quechua and Aymara, war dead were sometimes mummified, retained within the community, and revered during seasonal festivals.[56] These ancestors were not "gone" but were present, consulted on agriculture, conflict, and family well-being. Their continued presence ensured spiritual balance and collective guidance.

Among Māori of Aotearoa (New Zealand), the warrior dead were returned to their *iwi* (tribe) through the tangihanga ritual, a process of mourning, storytelling, song, and sacred protocol.[57] In some cases, bones were later exhumed and secreted away in hidden caves, ensuring their sacred mana would not be disturbed. The dead were not just mourned—they were cloaked in *tapu* (spiritual restriction) and honored as bearers of legacy.

Bones That Hold Power

Many Indigenous traditions hold that bones are not inert. They carry *mana*, power, presence. The Hawaiian *ali'i* (chiefs), for example, had their remains hidden in ocean cliffs or lava tubes to prevent desecration, as their bones were believed to retain divine authority. Mishandling a bone could bring imbalance or spiritual misfortune.[58]

This belief system contrasts sharply with Western forensic approaches, which often treat bones primarily as evidence. While forensic anthropology plays a vital role in identifying the war dead, especially in modern repatriation efforts, its practice on Indigenous remains has led to spiritual and ethical conflicts.

Efforts such as the Native American Graves Protection and Repatriation Act (NAGPRA) have attempted to restore the balance by facilitating the return of ancestral remains from museums and universities to tribal communities.[59] Yet many Indigenous leaders continue to advocate for deeper cultural sensitivity, consultation, and the restoration of ritual before scientific intervention.

The message is clear: to disturb the dead without ritual is to injure the living.

African Ancestral Return: The Dead Who Must Come Home

"A man who dies away from home is never truly dead. He waits."—A kan proverb, Ghana

Across the African continent, death is not departure—it is transition. And for warriors who die far from their ancestral lands, that transition remains incomplete until the body, or at least the spirit, returns home.

In many West African traditions, the dead join the ancestors only through proper burial rites performed in their homeland.[60] A body buried in foreign soil, without the correct rituals, becomes a wandering spirit caught between worlds, unable to take its place among the ancestors who guide and protect the living. This belief carries profound implications for warriors who die in distant battles.

Among the Akan peoples of Ghana and Côte d'Ivoire, death rituals are elaborate, communal, and essential. The body must be washed by specific family members, wrapped in cloth of particular colors, and buried with items that will accompany the spirit on its journey .[61] Funeral celebrations can last days, with drumming, dancing, and the recounting of the deceased's deeds. Without these rites, the dead cannot fully depart, and the living cannot fully grieve.

For African soldiers conscripted into colonial armies—fighting and dying in European wars, in trenches far from home—these rites were impossible. Hundreds of thousands of African troops served in World War I and World War II under British, French, Belgian, and other colonial commands.[62] Those who died were buried according to European military protocols with individual graves with standardized markers, or mass burials when conditions demanded. Their families received notifications—if they received anything at all—but no bodies returned.

The spiritual consequences within African cosmologies were devastating. Men who should have become honored ancestors instead remained suspended, their spirits unable to complete the journey. Communities that should have gathered to mourn, to celebrate, to release could do none of those things.

Among the Yoruba of Nigeria, the concept of *àṣà* (custom, tradition) governs how the dead must be treated.[63] Warriors were traditionally buried with their weapons and honored with festivals that recounted their bravery. Death far from home, without these rituals, was considered a form of spiritual violence—not just to the individual, but to the entire lineage.

The Zulu of South Africa held similar beliefs. A warrior's spirit required specific ceremonies to be released and welcomed among the ancestors.[64] During and after the Anglo-Zulu War and subsequent colonial conflicts, many Zulu dead were buried by British forces without regard for these traditions. The spiritual rupture persisted for generations.

Today, efforts to address this legacy are emerging. In some cases, soil from African soldiers' graves in France and Belgium has been ceremonially returned to their home communities, a symbolic repatriation when physical return is impossible.[65] Descendants have traveled to European cemeteries to perform belated rituals, speaking the names of ancestors who died a century ago.

These acts are not merely commemorative. They are restorative.

They acknowledge what colonial military practice refused to see: that African soldiers were not just labor or cannon fodder, but human beings embedded in spiritual systems that required specific care in death. To bury them without that care was not simply logistically convenient but a form of erasure.

The African perspective on war death teaches us that burial is not complete when the body is in the ground; it is complete when the soul is at rest. And for many African war dead, that rest has been delayed for generations.

They are still waiting to come home.

Asian Ancestor Veneration and the War Dead

"The dead are not gone. They eat with us. They watch over us. They wait for us to remember."—Vietnamese ancestral prayer

Across East and Southeast Asia, ancestor veneration forms the spiritual foundation of family and community life. The dead do not disappear—they remain present, active, and in need of care. And for those who die in war, that care carries particular urgency.

In Chinese tradition, death without proper burial is among the greatest misfortunes that can befall a person.[66] The soul of one who dies violently, far from home, or without appropriate rites becomes a *gui*—a hungry ghost, wandering and unsettled, capable of bringing misfortune to the living. The annual Hungry Ghost Festival exists, in part, to appease such spirits and offer them the care they were denied in death.[67]

For Chinese laborers who died on the Western Front during World War I—the nearly 140,000 men of the Chinese Labor Corps, thousands of whom perished from disease, accident, or the lingering effects of gas—these beliefs carried profound weight.[68] They were buried in France and Belgium according to Western protocols, with headstones bearing Chinese characters chosen by the British authorities. But the rituals that would have released their spirits—the burning of paper offerings, the ancestral tablets, the family ceremonies—never occurred.

Their spirits, within Chinese cosmology, remain unsettled.

In recent decades, descendants and community organizations have traveled to European cemeteries to perform belated ceremonies, burning incense, offering food, and speaking the names of great-great-grandfathers who died a century ago and a world away.[69] These acts are not symbolic—they are spiritually necessary.

Japanese traditions carry similar weight. The concept of *rei* (spirit) holds that the dead require ongoing care through memorial rites, particularly the annual *Obon* festival when ancestors return to visit the living.[70] Japanese soldiers who died in World War II on Pacific islands, in China, and in Southeast Asia numbered in the millions. Many were never recovered. Their families performed rituals with empty urns, burning incense before photographs, maintaining household altars (*butsudan*) that held no remains.

The Japanese government has conducted decades of recovery missions to bring remains home from former battlefields.[71] When bones are found, they are treated with profound reverence—washed, blessed, and returned to Japan for cremation and burial. The spiritual imperative is clear; the dead must come home, or they cannot rest.

Vietnamese ancestor veneration holds that the dead live on in the family altar, requiring regular offerings of food, incense, and prayer.[72] The Vietnam War created millions of deaths on all sides, many in circumstances that precluded proper burial. Mass graves, jungle burials, bodies lost to fire or flood. For Vietnamese families, the inability to perform proper rites created spiritual wounds that persist across generations.

After the war, the Vietnamese government and families undertook extensive efforts to locate and properly bury the war dead—including American remains, which were returned through diplomatic channels, and Vietnamese remains, which were sought through community networks and spiritual mediums.[73] The work continues today, driven not by politics but by the spiritual necessity of bringing the dead home.

For Asian American soldiers (Chinese Americans, Japanese Americans, Filipino Americans, Vietnamese Americans) who served and died in American wars, these traditions created particular complexities. Military burial protocols rarely accommodated Asian spiritual practices. Families received bodies prepared according to Western mortuary standards, without the rituals that Asian traditions required.

Some families adapted, performing traditional rites over graves already filled. Others maintained dual practices, with military honors followed by ancestral ceremonies. Many simply grieved within systems that did not see them.

The Asian perspective on war death reminds us that burial is not a single act but an ongoing relationship. The dead require care not just at the moment of interment, but across years, decades, generations. To abandon them to a single ceremony, however respectful, is to leave the work unfinished.

The ancestors are watching.

They are waiting.

And they remember whether we remembered them.

The Cost of Unritualized Death

War disrupts ancestral continuity. When Indigenous soldiers or civilians die far from their homelands or are buried without appropriate ceremony, the wound is not only individual. It is communal. During and after U.S. wars, Indigenous service members were often buried according to federal military standards, with little regard for tribal traditions. Families were left grieving with incomplete rites.

The act of burial, so central to spiritual health, became another space of cultural erasure.

In response, many Indigenous communities today are reasserting sovereignty over burial customs. Military funerals now sometimes include traditional songs, drumming, smudging, or clan symbols. Elders may bless remains before reburial; others fight for the repatriation of warrior bones from battlefields overseas.

> "When our warrior returns," said one Diné (Navajo) elder, "we must speak his name, cleanse his path, and tell the ancestors. Otherwise, he walks between worlds."[74]

Burial as Balance

To bury a warrior is not simply to mourn him. It is to rebalance the cosmos.

The land must be acknowledged. The spirit guided, the family supported, the story told. These acts are not optional in Indigenous traditions but vital expressions of relational care between the living, the dead, the land, and the spiritual world.

Western anthropology now increasingly recognizes the truth that burial practices are not just cultural artifacts. They are living systems of meaning, and they offer critical wisdom for how we approach death in war—not only with science, but with ceremony.

The Indigenous perspective teaches us that to care for the dead is to care for the living. That ritual is not just about belief. It is about continuity and wholeness.

When a warrior falls, he is not gone.He has merely changed form.He has returned to the circle.

SHARED DUST: RITUAL AS COMMON GROUND

"In the end, the earth does not care whose side you were on."—Veteran of the Iran–Iraq War

Ritual belongs to all of us.

Beneath the banners, beyond the battlelines, there is dirt. It holds no loyalty. It receives all equally.

War divides.But burial—the act of honoring the dead—can become a bridge.Throughout history, combatants have crossed enemy lines not for negotiation or surrender, but for something older and more sacred: the return of the dead.

TRUCE IN THE MUD

In December 1914, during World War I, something remarkable happened.

As frost coated the trenches along the Western Front, British and German soldiers began to sing. First in separate camps, then together. On Christmas Day, they emerged from their dugouts—not with rifles, but with open hands. They exchanged cigarettes, played football, and, most importantly, retrieved the bodies of their fallen.[75] In No Man's Land, enemies became mourners.

They dug graves side by side, removed the dead from barbed wire, and shared a silence far deeper than any treaty. It was not peace, but it was human.

Anthropologists call this ritual convergence—the spontaneous emergence of shared burial customs, even across cultural or linguistic barriers. It reflects the universal truth that when faced with death, people instinctively return to care.

Burial Amidst Guerrilla War

During the Vietnam War, this impulse reappeared in quieter ways.

Amid guerrilla warfare and fractured landscapes, American soldiers occasionally returned personal effects—photographs, letters, prayer beads—to local elders, who would bury them alongside the dead.[76] In some rare ceasefires, opposing forces silently agreed to retrieve casualties from the jungle floor. No words were exchanged. Only gestures.

The enemy may have worn a different uniform. But the body—bloody, still, sacred—was the same.

In the Middle East, similar moments occurred during the Iran–Iraq War. Soldiers from both sides met in desert corridors to exchange corpses, sometimes reciting verses from the Qur'an over enemy remains. In such acts, ritual overcame ideology. The need to honor the dead outweighed the desire to dehumanize the enemy.

Silent Acts, Unwritten Grace

These moments are not often captured in official histories. But they exist in the dust under a soldier's fingernails, in the improvised burial mound marked with a helmet, in the whispered prayer said in a language the mourner does not speak.

In Ukraine, both sides have exchanged dead through negotiated channels or silent truces, even while hostilities raged. In Syria, local recovery teams and NGOs have organized efforts to return bodies across militant lines, motivated not by diplomacy, but dignity.

These gestures are not grand. They are mundane acts of grace.

A chaplain pausing over an unknown fighter.A medic removing a wedding band and laying it on a chest.A gravedigger folding a blanket over an enemy's body, simply because it's cold.

They are not political.They are profoundly human.

Ritual Beyond Flags

Whether Christian or Muslim, Buddhist or animist, soldier or civilian, the act of burial holds a similar rhythm:

- Pause
- Wash or wrap the body
- Mark the site
- Say a name—or acknowledge that one is missing

This commonality matters.

Even in the fiercest wars, burial remains one of the last places where shared values emerge. Where enemies lower their voices. Where even those who cannot forgive still choose to remember.

Because in the end, we are more than what we fight for.We are what we bury.

The Sacred Task: Religious and Spiritual Care in Battlefield Burials

"Burying the dead is a holy act. Not just for the faithful—but for the human."—Military chaplain, Korea

Long before dog tags, field kits, or casualty reports, there was sacr edness.The care of the dead has always been more than procedural—it is spiritual labor.

Across cultures and centuries, the battlefield has been not only a place of destruction but of reverence. Where guns fall silent, prayer often begins. Not for strategy, not for victory, but for the dignity of the fallen.

Chaplaincy at the Front

From the American Revolution to Iraq and Afghanistan, military chaplains have walked alongside soldiers—not only to offer spiritual guidance in life, but to stand at the threshold of death.

During World War I, chaplains held services in trenches, prayed over mass graves, and wrote letters to families of the fallen. During World War II, they crawled across beaches with medics, delivering last rites under fire. In Vietnam, they rode in helicopters to remote firebases, carrying Bibles, Qur'ans, rosaries, or sometimes just silence.

They belonged to every faith—Christian, Jewish, Muslim, Buddhist, Indigenous. And though their words differed, their presence did not. They stood in for families who could not be there. They said what needed to be said. They bore witness.

"I didn't always know their names," said one WWII chaplain. "But I never let them be buried without a word."[77]

Faith Across Lines

Burial care has never belonged to one religion.

- Muslim troops have been buried facing Mecca, with appropriate ablutions performed when possible.
- Jewish soldiers have been interred with Stars of David, accompanied by prayers for the soul's peace.

- Native American service members have received tobacco offerings, clan songs, or the blessing of the Four Directions.
- Buddhist nurses have whispered mantras.
- Atheist soldiers have been honored with silence—and the memories spoken aloud by their comrades.

These are not mere formalities. They are rites of passage, moments that acknowledge that even amid chaos, we remember how to care.

When the Rites Cannot Be Done

Sometimes, there is no chaplain. No book. No time.But still, someone tries.

A soldier mutters the Lord's Prayer while closing a friend's eyes.A Marine quietly places a rosary in a body bag.A field medic—agnostic—folds hands and covers a face, not out of belief, but out of instinct.

These are not acts of religion. They are acts of humanity.

And they remind us that sacredness is not dependent on belief. It is carried in the gesture. The pause. The presence.

Burial Without Faith – Still Sacred

Even among the nonreligious, burial takes on ritual form:

- The folding of a flag with crisp corners
- The reading of a name at roll call
- The return of a boot to a grieving mother's porch
- The placing of a stone at a grave's edge

These are acts of remembrance, spiritual not because they invoke a god, but because they affirm the value of a life. Anthropologists note that ritual, not belief, is what defines sacred practice.[78] It is the repetition of gestures imbued with care, meant to restore order in the face of death.

The battlefield is not a temple. But sometimes, it becomes one.

A shallow grave dug in silence.A battlefield cross.A whisper into the dust.

These are the altars of war. And those who kneel there—chaplains, gravediggers, medics, friends—become, for a moment, sacred caretakers.

They do not just bury the dead.They remember them.

And in doing so, they remind us who we are.

To Lay Down the Sword

There is a moment in every war when the sword is lowered—not in surrender, but in reverence. It does not make headlines. It is not recorded in official logs. But it is felt.

It happens when a soldier bends over a fallen comrade—or even an enemy—and reaches, instinctively, to close the eyes. It happens when hands, calloused from battle, pause to place a stone, a flower, a name upon the earth. It happens when silence falls, not from ceasefire, but from something deeper: the recognition of humanity.

These gestures are older than politics, older than uniforms, older than war itself.

Anthropology teaches us that ritual is not decorative—it is foundational. In the chaos of violence, it is ritual that restores meaning. It is how we make sense of loss. How we mark the boundary between death and forgetting. How we stay human.

To bury the dead is not merely to dispose of a body. It is to say: *this person mattered. This life belonged to us. We will carry them, even now.*

Across deserts and forests, battlefields and oceans, humans have always done this. Whether with scaffold or scroll, smoke or psalm, cotton shroud or military flag, we bury to remember, and in remembering, we rebuild.

Even today, when wars are fought with drones and digital screens, when a name can be retrieved with a single swab of DNA, there is still the need to pause. To hold the body. To give it shape and place and finality.

From scaffold burials on the Great Plains to prayer fires lit in Pacific caves...From battlefield interments in Flanders to genetic testing in sterile labs...From whispered prayers in foxholes to midnight repatriations under flags flown at half-staff...The impulse remains the same: to carry the dead back to belonging.

This is the anthropology of war death. Not a study of violence, but of what we do after it. Not a glorification of suffering, but a tribute to the sacredness of response.

Because when the body is lowered into the earth—wrapped in cloth, carried in casket, laid in sand—it becomes more than remains; it becomes a threshold.

A place where war ends, and memory begins.

In that moment, we are not conquerors. We are not casualties.We are caretakers.

And to lay down the sword—just for that moment—is not weakness. It is strength of another kind. The kind that mourns, that honors, that heals.

The kind that tells the dead:You are known.You are not lost.You are brought home.

What we do for the dead shapes what we become among the living.It is the first step in peace. The first word of truth. The first breath of reconciliation.

In the end, there is always the grave. And it waits not as an ending, but as an invitation to remember who we are. To become more than what we've survived. To lay down the swordand pick up the burden of love.

Spotlight: Dr. Clyde Snow – The Father of Forensic Humanitarianism

"The bones don't lie."—Dr. Clyde Snow

In the history of forensic anthropology, few figures loom as large as Clyde Collins Snow. A cowboy-booted Texan with a gift for translating scientific findings into plain language, Snow did more than anyone to establish the principle that forensic science could—and should—serve the cause of human rights.[79]

Snow's early career was conventional. He worked at the Federal Aviation Administration's Civil Aeromedical Institute, consulting on aircraft disasters and identifying victims of crashes. But in 1984, he received an invitation that would transform both his career and the field. The newly democratic government of Argentina asked him to help identify the remains of the "disappeared"—the thousands of citizens kidnapped, tortured, and murdered by the military junta between 1976 and 1983.[80]

What Snow found in the mass graves of Argentina was not just bone—it was evidence. Evidence of torture. Evidence of execution. Evidence of crimes that the perpetrators had tried to erase. The bones,

Snow demonstrated, could testify when witnesses had been silenced.

He trained a generation of Argentine forensic anthropologists—the *Equipo Argentino de Antropología Forense* (EAAF)—who would continue the work and export the methodology worldwide.[81] His students went on to investigate mass graves in Guatemala, El Salvador, Bosnia, Rwanda, and beyond. The discipline Snow helped create has since identified thousands of victims and provided evidence for war crimes tribunals around the world.

Snow's approach was distinctive. He treated every set of remains as a person deserving of respect, not merely a specimen for analysis. He spoke to the dead as he worked, acknowledging their presence, honoring their suffering. "I do believe that the bones talk to me," he said. "Not in a creepy way. But they tell me their secrets."[82]

He testified at the trial of Argentine junta leaders. He examined the remains of Josef Mengele. He identified victims of John Wayne Gacy. But his greatest legacy was demonstrating that forensic science could serve justice, and that the same techniques used to identify air crash victims could expose the crimes of dictatorships.

Snow died in 2014, but his influence endures. The forensic teams now working for the DPAA, the ICRC, and human rights organizations worldwide operate within frameworks he helped establish. The principle that the dead deserve identification—that bones tell stories that must be heard—is his most lasting contribution.

"Of all the forms of murder," Snow once said, "state-sponsored killing is the most barbaric. Because the state has the resources to make bodies disappear." His life's work was proving that disappearance is never complete. The bones remain. And with patience, skill, and commitment, they can speak.[83]

He gave the dead a voice.

And he taught a generation to listen.

Artifact: The Clay Token from Tell Banat: An Ancient Burial Receipt

Unearthed in modern Syria at the Bronze Age site of Tell Banat, archaeologists found small clay disks—some etched with marks, others shaped like miniatures of human figures. These were discovered in mass graves and burial mounds, dating back over 4,000 years.

Researchers believe these were burial receipts, tokens offered in acknowledgment that someone had been buried, that their death had been marked. In a time before writing was widespread, these tokens served as ritual documentation. They did not list names, but they bore intent, acknowledging that a life existed, ended, and was remembered.

These ancient disks, smaller than a coin, reveal something profound: even in prehistoric war zones, people paused to name the dead—not with words, but with symbols of witness.

They are the ancestors of dog tags, field ledgers, and memorial plaques. They are echoes of that enduring question: *who lies here?*

And they remind us that the instinct to honor—even in the simplest of gestures—has always been part of what makes us human.

Interlude: Before the Dust Settles

"The war wasn't over when the shooting stopped. It was over when the last body was laid to rest."—Mortuary Affairs Officer, Afghanistan, 2012

They come after the headlines fade. After the troops withdraw, after the battlefield is silent but still filled with bones.

They come in khaki, or lab coats, or plain clothes and gloves.For ensic teams. Chaplains. Volunteers. Family liaisons.Not to fight—but to gather. To tend. To bring home.

They kneel in craters. Lift rusted dog tags. Match teeth to photographs. Read trauma in femurs like verses in a sacred book.

They do not come for glory; they come because someone must.

Before Chapter Nine begins—before we turn to the modern wars that blur body and machine, that challenge remembrance with digital detachment—this interlude holds a breath.

A pause to remember that the care of the dead is not a matter of era. It is a matter of heart.

Whether Bronze Age or drone age, we bury the fallen to hold ourselves together.

Before we march forward, let us stand once more beside the grave and remember who we are.

Reflection

Clara Barton's work continues. And I am blessed to be a part of it. In addition to my Red Cross Service to Armed Forces volunteer work, I volunteer with Women of the World, an organization serving refugee and asylum-seeking women in my community.

Many of these women have been forcibly displaced by the same conflicts that fill today's headlines and the dozens of other crises that generate refugees but rarely sustained attention.

Through Restoring Family Links—a program that traces its roots to Clara Barton's Civil War work and Henri Dunant's vision at Solferino—I help these women search for family members scattered by war and displacement. A mother looking for a son detained in Syria. A wife searching for a husband who disappeared during the journey to safety. A daughter trying to confirm whether her father is alive or dead.

The work is slow. The answers are often incomplete. Sometimes there are no answers at all. But the searching matters. The act of looking—of refusing to accept that someone can simply disappear—is itself a form of care. It tells the missing they are not forgotten. It tells the family their grief is witnessed.

This next chapter examines modern warfare and its new forms of invisibility—drones that kill without presence, systems that process death as data, and technologies that create distance between the living and the dead.

But it also affirms that care persists.

That someone is still doing this work. That the thread connecting Barton and Dunant to the present day remains unbroken. I hold a small piece of that thread. It is an honor I do not take lightly. The dead are still being buried. The missing are still being sought. The families are still waiting for answers.

And someone must still do the work.

Chapter Nine

Modern Wars and the Evolving Dead: Iraq, Afghanistan, and Beyond

"In this war, the dead come home in real time. But we still don't know how to carry them."—U.S. Mortuary Affairs, NCO, 2006.

The wars in Iraq and Afghanistan were not the deadliest American conflicts, but they were among the most intimate. Fought in the age of smartphones, satellites, and livestreamed combat, these wars transformed not only how soldiers died but how they were remembered, mourned, and forgotten.

In earlier wars, weeks or months might pass before news of a death reached home. In these modern conflicts, a name, a photo, and a flag-draped coffin could appear online within hours. The dead now returned under the eye of the public—but often without the understanding of the nation they served.

This chapter explores how war death has evolved in the 21st century. It examines the rise of digital mourning, the reorganization of Mortuary Affairs units, the role of private contractors and noncombat personnel in body recovery, and the shifting rituals of return in an age when war is constant but rarely discussed.

But it also exposes a deeper tension: that even as our technologies improve, our ability to honor the dead—and those who carry them—has not kept pace.

We ask what it means to bear witness in a war that feels far away.

We look at how grief is processed under surveillance.And we hear from those who continue to bury the fallen, even as the nation tries to move on.

In these pages, we meet the modern death workers—those who walk beside the casket, who close the bags, who catalog the remains, who fold thc flag. And we remember that though the war has changed, the weight of the dead has not.

Interlude: The Dust Still Clings

"You come home with the sand still in your boots. And the names still on your hands."— Army Mortuary Affairs Specialist, Iraq, 2007

They say these wars were cleaner. That they used precision, drones, data. That the battlefield was somewhere else—always *over there.*

But those who handled the dead know different.

They remember the heat shimmering off a black bag on the tarmac. The scent of blood in a Humvee. The sound of a mother's scream through a satellite phone.

They carried limbs. Held dog tags. Zipped the silence.

And when they came home, no one asked what they had seen. No one asked what they still carried.

Their grief was not televised. Their hands were not folded in parades. Their names are not listed in bronze.

But the dust still clings to them. The dust of war, the dust of memory, the dust of the ones they could not save.

They were not heroes.They were not martyrs.They were caretakers.

And still—they are the reason the fallen made it home.

Death at a Distance: The Drone Age and the Disappeared Body

"I watched him die on a screen. Then I drove home and picked up my kids from school."—MQ-9 Reaper sensor operator, 2014

In the 2st century, death went remote.

For most of human history, killing and burial were intimately connected. The warrior who struck the blow often saw the body fall. Someone—sometimes the killer himself—would address the remains. But modern warfare has severed that connection. Today, a pilot in Nevada can launch a Hellfire missile at a compound in Yemen, watch the explosion on a monitor, and never see what—or who—emerges from the rubble.[1]

The drone age has created a new category of war dead: the killed-but-unseen. These are combatants, suspected combatants, and civilians whose deaths are recorded by satellite imagery and sensor feeds, but whose bodies are never recovered by the forces that killed them. They become data points in strike assessments, not individuals requiring burial.

For the communities where these strikes occur, the consequences are devastating. Families search through rubble for remains. Bodies are buried hastily, sometimes in pieces, often without the rituals that tradition requires.[2] The dead are mourned locally, but internationally they exist only as statistics—"military-age males" in after-action reports, numbers in classified tallies that may never be fully disclosed.

The psychological distance this creates extends to the operators themselves. Studies of drone pilots and sensor operators have documented a distinctive form of moral injury: the trauma of killing without presence.[3] They watch their targets for days or weeks before strikes, learning their patterns, sometimes seeing their children. Then they press a button, watch the explosion, and the screen goes blank. There is no body to recover. No grave to mark. No ritual to perform.

One sensor operator described the experience: "You see everything. The heat signatures cooling. The people running. And then you log off and go to lunch. It's the most intimate killing imaginable, and also the most disconnected."[4]

For the military mortuary system, drone warfare creates a gap. When American forces conducted ground operations in Iraq and Afghanistan, bodies—enemy and civilian—were encountered, documented, sometimes buried. But strikes conducted from altitude, without ground forces present, leave no such accounting. The dead become data before they become bodies. And often, they never become bodies at all. At least not to those who killed them.

This is not to say that drone warfare is uniquely immoral. Every form of warfare creates distance between killer and killed through artillery, airstrikes, naval bombardment. But the drone age has made that distance both more extreme and more intimate. Operators see faces. They watch daily routines. And then they end lives without ever smelling smoke, hearing screams, or confronting remains.

The burial implications are profound. In communities across Yemen, Pakistan, Somalia, and Syria, families have developed their own rituals for drone-strike victims—ways of mourning the dead whose bodies may be incomplete, contaminated, or lost entirely.[5] These are not military funerals; there are no flags, no honors, no recorded names. There is only grief, improvised and invisible to the forces that caused it.

For those who operate the drones, there is often no ritual at all. The mission ends, and another mission begins. The dead are filed away in classified databases.

But they are still dead.

And someone, somewhere, is still mourning them.

The drone age asks a question that earlier chapters have not confronted: what happens to burial when there is no body to recover? What happens to the caretakers when the killing is done from a continent away?

The answers are still forming. Some drone operators have sought chaplain support, therapy, or informal rituals of their own as ways of acknowledging the weight of what they've done.[6] Some have left the service, unable to reconcile the intimacy of watching with the distance of killing. Others carry the images in silence, scrolling through faces they will never forget.

The dead at a distance are still dead.

They still deserve to be mourned.

But in the drone age, the question of who mourns them—and how—remains unanswered.

Burial at Sea Today

The practice of burial at sea continues in the modern military—not just for deaths at sea, but as an option for any eligible veteran.

The U.S. Navy performs burial at sea services for active-duty personnel, retirees, and veterans who request it. The ceremony follows the same essential form it has for centuries: the body or cremated remains are prepared, prayers are offered, and the dead are committed to the deep. The coordinates of the burial are recorded and provided to the family.[7]

For some families, especially those whose loved ones served in the Navy or felt a deep connection to the ocean, burial at sea offers a fitting tribute. For others, the absence of a physical grave is too much to bear. They need a place to go, a stone to touch, a location where memory can be anchored.

The choice reflects a fundamental tension in how we remember—the tension between presence and absence, between the need for sacred ground and the recognition that the dead do not stay in one place.

The sea has always taken the dead. And the sea does not give them back.

But it holds them. And for some, that is enough.

THE DIGITIZED DEAD: WARFARE AND BURIAL IN THE INFORMATION AGE

"I found out on Facebook that he'd been killed. The Army hadn't even called yet."—Sister of fallen Marine, 2005.

In modern war, death travels faster than grief can comprehend.

What once took weeks to reach a family's doorstep now arrives in a matter of hours—through push notifications, tagged photos, or a sudden change in someone's profile picture. In Iraq and Afghanistan, the death of a service member could be streamed, posted, or whispered into the wrong inbox before the official casualty team ever rang the

doorbell.

The age of digitized death ushered in the haunting contradiction of real-time information with delayed meaning. The moment of death—once held in sacred delay—was now a spectacle. Helmet cams recorded the final breath. Drones surveyed the aftermath. A mother in Missouri might learn of her son's death from a Marine's roommate's Instagram post, blurred and cryptic, yet unmistakable.

For Mortuary Affairs teams, the burden of care grew heavier—not because the dead were more numerous, but because they were more visible. The work of preparing, identifying, and returning the dead unfolded beneath a digital microscope. One misstep—a misspelled name, a delayed transfer, a leaked photo—could ignite public outrage and deepen a family's grief.

And yet, even with the weight of surveillance, there was silence around the act itself. Most Americans scrolling through casualty headlines never paused to ask: *who picked him up? Who zipped the bag? Who held his hand until the chopper came?*

This was a new kind of invisibility born not of secrecy, but of saturation.

Death became a data point to be collected, categorized, and stored in databases of biometrics. Forensic identification now relied on DNA, dental scans, and digital fingerprints. In the 2000s, the U.S. military established a DNA repository for all active-duty personnel, ensuring that even partial remains could be named. Precision improved. Closure became more possible.

But at what cost?

For those in the field, the intimacy of death now came with the detachment of data. What had once been a sacred act of recognition—the naming of the dead—became a series of screen clicks and autopsy codes. And still, the burden of reverence remained. Someone had to stand over the body. Someone had to weep.

Grief, too, evolved.

Gold Star families now mourned in public spaces. Funeral livestreams reached distant friends. Tribute pages gathered thousands of comments—some heartfelt, others cruel. Memorials lived in hashtags. Mourning became social. But not always sacred.

In this age, ritual was harder to hold. It flickered through screens, it vanished beneath algorithms, it became a page you could scroll past.

But somewhere—always—someone still carried the body.Stil l whispered the name.Still bore the weight.

Death in the information age may travel through fiber optics.But it still arrives by hand.

Spotlight: The Final Salute at Dover

Dover Air Force Base is more than a military installation. It is a threshold.

Since 2001, every fallen American service member from Iraq and Afghanistan has returned through Dover, met not by noise or ceremony, but by quiet choreography. The Dignified Transfer Ceremony is not a press event. It is not a eulogy. It is a sacred act performed in silence, boots moving in perfect rhythm beside the flag-draped casket.

For years, these ceremonies were closed to the public. But in 2009, under President Obama, families were given the choice to allow media coverage. Some accepted, believing visibility mattered. Others declined, craving privacy.

Each casket carried a world. Each fold of the flag a wound. And each moment of stillness at Dover was a whispered vow: *you are home.*

You are not forgotten.

Artifact: A Thumb Drive from Fallujah

In 2006, a Marine mortuary team found a thumb drive on the body of a fallen soldier outside Fallujah. It was not military issued. It bore no insignia. But inside were photos: a birthday cake, a baby's first steps, a letter half-written and saved.

The drive was cleaned, restored, and returned to the family.Not for intelligence.But for memory.

In the digital age, our identities no longer live only in tags or uniforms. They live in code. In pixels. In pieces of ourselves saved in silence.

That drive was not just data; it was a man's life.And someone cared enough to carry it back.

Body, Bag, Database: The Modern Mortuary System

"We're not returning remains. We're returning identity."—Air Force Mortuary Affairs Operations (AFMOA) team member

In ancient wars, the dead were carried by hand, buried by kin, and remembered in story.In modern war, the dead are flown, coded, cataloged. Still, they are remembered.But now, the memory comes wrapped in protocol.

The journey home for today's fallen begins in dust and fragments, often in terrain where no ritual can yet be performed. Mortuary Affairs specialists retrieve the remains under duress, amid insurgent gunfire, smoldering wreckage, or craters left by IEDs. Some arrive whole. Some arrive in parts. All are received as sacred.

From there, a chain of reverent logistics begins.

Remains are evacuated through secured corridors and refrigerated aircraft, each tagged with biometric identifiers, time stamps, and GPS logs. A dedicated chain of custody ensures nothing is lost—not even a wedding ring, a half-scorched letter, a zippered uniform pocket. But behind the paperwork, there is always a person. And someone always knows it.

At the center of this sacred system is Dover Air Force Base in Delaware, home of the Air Force Mortuary Affairs Operations (AFMAO). There, behind sealed doors and silent corridors, the dead are prepared for return with painstaking care. Dover is not a factory. It is a sanctuary disguised as infrastructure.

Upon arrival, the remains pass through multiple hands and disciplines: forensic anthropologists, embalmers, autopsy pathologists, grief counselors, chaplains, and uniform coordinators. They work not just to restore the body but to restore the person.

Each uniform is steamed. Each ribbon placed by hand. Hair is trimmed, if present. Cosmetics are gently applied. Hands are sometimes positioned holding a rosary or left open as if in peace.

The goal is dignity—not only for the family, but for the dead.

Where WWII morticians worked with guesswork and manual charts, today's teams utilize digital dental scans, DNA comparisons, and biometric analysis. Identity is triangulated with unmatched precision using medical records, antemortem imaging, and in rare cases, mitochondrial DNA from maternal relatives.

Yet even in this era of medical certainty, the work is never purely technical.

Each body receives a pause.A moment of prayer.A breath of reverence.

Then comes the Dignified Transfer Ceremony. Under solemn watch, the casket is carried by a uniformed detail onto the tarmac. There are no anthems, no words. Only boots on concrete and silence as deep as grief.

The mission doesn't end at Dover. In many cases, a military escort—often someone of the same branch or unit—accompanies the body all the way home. Commercial flights. Layovers. Cargo holds. Always seated nearby, always watching. Escorts eat beside the casket, sleep near it, pray over it.

Their job is to ensure the dead are never alone.

Modern mortuary systems also include psychologists, case workers, and funeral liaisons. Yet the emotional toll remains enormous. Mortuary Affairs personnel report high rates of depression, PTSD, and moral injury. They carry not only the body, but the weight of national remembrance—and the silence that follows when the flags are lowered.

In this system, the body becomes both relic and record. A sacred text that must be delivered intact.

And behind every digital file, every DNA scan, every dress uniform folded perfectly, there remains one truth:It is still someone's son.Someone's daughter.Someone who should not have had to die.

DPAA: The Past as Present Work

"We're not just finding bones. We're keeping a promise that's seventy years old." —DPAA recovery team leader, 2022

While the wars of today produce new casualties, the wars of the past are not finished.

Every week, the Defense POW/MIA Accounting Agency announces new identifications—names returned to families who have waited decades. A Marine lost at Tarawa in 1943. A soldier missing from the Chosin Reservoir since 1950. A pilot shot down over Laos in 1967. Each announcement represents years of research, excavation, and forensic analysis. Each represents a promise kept.[8]

The DPAA is the institutional embodiment of a commitment that no other nation maintains at this scale: that the missing will be sought, no matter how long it takes. More than 81,500 Americans remain unaccounted for from past conflicts.[9] The majority—over 72,000—are from World War II, their remains scattered across Pacific islands, European battlefields, and ocean depths. Another 7,500 are missing from Korea. Approximately 1,580 from Vietnam.

The work is painstaking. A single identification can require years of effort. Researchers comb through archival records, declassified documents, and eyewitness accounts to locate probable burial sites.

Field teams—archaeologists, anthropologists, explosive ordnance specialists—conduct excavations in jungles, deserts, and former battlefields that may still hold unexploded munitions.[10] The remains they recover are transported to laboratories in Hawaii and Nebraska, where forensic specialists analyze every fragment.

DNA technology has transformed what is possible. Mitochondrial DNA, extracted from degraded bone, can be compared against reference samples provided by family members—sometimes by grandchildren or great-nieces who never knew the missing.[11] What was once scientifically impossible is now routine. Names that seemed lost forever are being recovered.

In 2023 alone, the DPAA identified the remains of 186 service members.[12] Each identification triggered a notification, a phone call to a family that had stopped expecting one. Each led to a funeral, often with military honors, for someone who died before many of the mourners were born.

The families who receive these calls describe the experience as both ending and beginning. The uncertainty is over. The questions are answered. But now comes the grief that was suspended, the funeral that was postponed for fifty, sixty, seventy years.

The DPAA's work intersects with the international community. Recovery operations require diplomatic coordination with foreign governments—some cooperative, some resistant. Excavations in Vietnam and Laos continue through bilateral agreements.[13] Access to North Korea remains largely impossible, though brief windows have occasionally allowed recovery missions. Each site recovered represents not just forensic work, but years of diplomatic effort.

The men and women who perform this work are caretakers in the truest sense. They handle remains with the same reverence that GRS soldiers showed in 1918, that Mortuary Affairs specialists showed in 1968. They speak the names. They document everything. They ensure that the dead are not abandoned to anonymity.

The past is not past.

It is present work.

And it continues today.

Spotlight: AFMAO's Honor Guard: The Keepers of Dignity

At Dover, a select group of military personnel make up the Mortuary Affairs Honor Guard, a team trained to execute every movement of the Dignified Transfer Ceremony with precision and reverence. They are not assigned this duty casually. Each member is evaluated not only on discipline but emotional steadiness, empathy, and silent strength.

They carry the casket, fold the flag, bow their heads.But they also grieve.

Some carry the memory of friends they've transferred. Some recite quiet prayers—unspoken and unseen. Their work is filmed, streamed, immortalized. And yet, most Americans don't know their names.

They are the invisible stewards of our national promise: that the dead will be received with honor, no matter what.

Artifact: The Personal Effects Pouch

Among the most emotionally charged items returned to families are the personal effects of the fallen.

At Dover, these belongings are meticulously cleaned, cataloged, and packed in what is known as a Personal Effects Pouch. Inside, one might find:

- A torn photo of a child
- A wedding ring threaded with soot
- A lucky coin carried since boot camp
- A bloodstained prayer card or field notebook

These are not inventory. They are ghosts in fabric.

In one case, a soldier's watch was stopped at the time of the blast. It was returned to his mother, who still winds it once a year on the anniversary of his death.

Objects like these remind us that remembrance is not abstract.It is worn. Held. Wept over.

They are not just remnants. They are *reentry points for memory.*

War's New Witness: Contractors, Medica, and Invisible Labor

"No one thinks about who pick up the pieces. We just do it."—Civilian contractor, Iraq

In today's wars, the front line is blurred—and the body bearers wear many uniforms.

Gone are the days when war death was handled solely by military personnel in official capacities. The post-9/11 conflicts in Iraq and Afghanistan created sprawling, hybrid operations involving soldiers, medics, contractors, local allies, interpreters, and civilian logisticians. In this tangled web of war, death does not discriminate, and neither does the labor of carrying it.

Civilian contractors, often hired for roles ranging from transport to security, increasingly found themselves tasked with battlefield recovery. Some had prior military experience. Many did not. In the wake of IED blasts or ambushes, they were the ones on site—loading remains, retrieving scattered limbs, guarding convoys that contained the dead.

There was no script. No chaplain. No command support. Just the raw, immediate task: *get the body. Protect the memory.*

Unlike formal Mortuary Affairs teams, contractors lacked the structured debriefs, psychological support, or ceremonial frameworks available to enlisted service members. Their trauma became privatize, felt deeply, but rarely witnessed—let alone honored. When their contracts ended, they returned to civilian life with war inside them and few places to speak of it.

Medics and corpsmen, too, became caretakers of the fallen. Trained to save lives, they often ended up preparing the dead by stripping gear, writing down names, closing eyes. Many kept notebooks of the ones they lost. Some recited prayers they barely remembered. Others folded the hands of strangers simply because *someone should*.

And then there were the local allies: Afghans, Iraqis, Kurds, interpreters, drivers. These individuals were the first on the scene, often pulling bodies from vehicles or roadside blasts. They translated death for American forces, identifying clothing, gestures, names. Many risked their lives to retrieve remains that others would not approach. And when the war ended, many were abandoned, denied asylum, or targeted for their collaboration.

Their contributions were profound. Their trauma was left unwritten.

The burden of remembrance was also borne by chaplains, dog handlers, and even drone analysts—those who never touched the body but saw it first, or last. One handler buried her fallen military dog beside her own father. One chaplain said a prayer over a Syrian fighter because there was no one else to speak his name.

This is war's new witness: fragmented, unofficial, uncelebrated.

The dead are still carried.But now, the carriers are many.And most are never seen.

Syria, Ukraine, and the Wars of Right Now

"We bury them where we can. Sometimes in gardens. Sometimes in parks. Anywhere the shelling has stopped long enough."—Volunteer recovery worker, Mariupol, 2022

As you read these words, burial work is happening.

In Ukraine, volunteers in yellow vests exhume bodies from mass graves, photograph faces, collect DNA samples, and search for identification—anything that might connect an anonymous corpse to a waiting family.[14] In Syria, recovery teams navigate between armed factions to retrieve remains from collapsed buildings, prison sites, and unmarked burial grounds that hold the dead of more than a decade of civil war.[15] In Gaza, families dig through rubble with their hands, searching for children whose bodies may never be found.

These are the wars of right now. They are producing mass casualties, mass graves, and profound disruptions to the rituals that give death meaning. And in each of them, someone is doing the work this book has traced across centuries: recovering the dead, identifying them, returning them to those who loved them.

The scale of loss in contemporary conflicts defies comprehension. Syria's civil war has killed over 500,000 people; tens of thousands are still missing.[16] Russia's invasion of Ukraine has produced mass graves in Bucha, Izium, and Mariupol—sites that have become synonymous with atrocity.[17] The dead are combatants and civilians, soldiers and children, identifiable and anonymous.

In these contexts, traditional burial rituals are often impossible. Shelling makes cemeteries inaccessible. Siege conditions prevent proper preparation of bodies. Religious requirements—the Muslim obligation to bury within twenty-four hours, the Orthodox Christian tradition of open-casket viewing—cannot always be fulfilled.[18] Families improvise, adapting rituals to impossible circumstances, burying loved ones in backyards and public parks, marking graves with whatever materials survive.

The documentation work is urgent and dangerous. In Ukraine, prosecutors and forensic teams work alongside military operations to collect evidence of war crimes by photographing bodies, recording injuries, preserving DNA.[19] This is burial work fused with justice work: the dead as witnesses, their bodies as testimony. The International Criminal Court has opened investigations. The evidence being gathered now will shape trials that may occur years or decades in the future.

International organizations play crucial roles. The International Committee of the Red Cross has maintained operations in Syria throughout the conflict, documenting detention, facilitating prisoner exchanges, and supporting families seeking information about missing relatives.[20] In Ukraine, the ICRC has assisted with body recovery and identification, drawing on expertise developed across decades of conflict response.[21]

But the work exceeds institutional capacity. Much of it falls to volunteers—local people who cannot bear to leave their neighbors unburied. They are not trained forensic anthropologists. They are teachers, mechanics, retirees, students. They learn as they work, developing protocols through necessity, finding ways to preserve dignity when everything else has been destroyed.

One Ukrainian volunteer, interviewed in 2023, described the work: "We find them in basements, in cars, in the street. Some have documents. Some have nothing. We photograph everything. We take samples. We wrap them carefully. And we bury them with a number, hoping someday the number becomes a name."[22]

This is the same work that GRS soldiers performed in France in 1918, that Mortuary Affairs specialists performed in Vietnam in 1968, that DPAA teams perform today on Pacific islands. The tools have changed—DNA kits instead of dental charts, digital databases instead of paper ledgers—but the essential task of ensuring that the dead are not lost, that families receive answers, and that memory survives chaos remains.

The wars of right now will produce their own generation of caretakers marked by what they have seen, carrying trauma that may take decades to fully express. Some will be recognized. Most will not. They will return to ordinary lives, if they survive, holding memories that their neighbors cannot imagine.

They are the witnesses.

They are the ones who stayed.

And they are working right now, while you read this page.

Spotlight: The Civilian Mortuary Tech

In 2011, at a forward operating base in Kandahar, a civilian mortuary technician known only as "Ms. Rose" handled the remains of fifty-seven U.S. and coalition soldiers over one six-month period. She was not military. Not clergy. She had no rank.

But every time a body came in, she lit a small candle.

She would speak the name—when known—and lay a folded prayer slip beneath the foot. Not for religion but for ritual.

She wore gloves to preserve evidence.But she always touched the hand, bare—just once.So the dead knew they were not alone.

Artifact: The Notebook of a Corpsman

Found in a storage trunk at Camp Pendleton, a Navy corpsman's leather-bound notebook included over 60 names.

Each one was handwritten.Some had stars beside them.Some had short lines: "sang in church," "wrote poems," "brought me tea once."

At the back, a final page read:

"They didn't die nameless. Not to me."

The corpsman never published this journal. Never mentioned it in his VA intake forms.

But he kept it.And that is a kind of memorial, too.

The Red Cross in Modern Conflict: Service to Armed Forces Today

The work that Clara Barton began continues in the present tense.

Today, the American Red Cross Service to Armed Forces (SAF) program provides emergency communication services, connecting military families with their loved ones during crises.[23] When a service member's parent dies, when a child is born, or when a family emergency requires immediate notification, the Red Cross facilitates communication across war zones and training deployments, ensuring that service members can be reached and families can be supported. The organization's Restoring Family Links program continues its tracing mission, connecting families separated by conflict and disaster around the world.[24] In Syria, in Ukraine, in Yemen, families submit inquiries seeking news of missing relatives. Red Cross and Red Crescent workers search databases, contact detention facilities, and coordinate with other national societies to provide answers when possible—or to document the search when answers cannot be found.

These are not historical functions. They are happening now.

In 2023, the American Red Cross responded to thousands of emergency communications involving military families.[25] Caseworkers helped navigate the bureaucracy of deployment, the confusion of casualty notification, and the long aftermath of loss. They served families whose needs did not end when the funeral ended—families navigating benefits, struggling with grief, searching for community.

For those of us who serve in SAF and RFL capacities—as educators, volunteers, caseworkers—the work carries weight. We stand at the intersection of military and civilian life, helping families bridge the gap between deployment and home. We do not handle remains. We do not dig graves. But we carry messages. We connect the separated. We witness the particular grief of military families and try to offer what support we can.[26]

This is not the same as Mortuary Affairs work. But it is part of the same continuum: the recognition that war's aftermath extends far beyond the battlefield, that families need support, and that care must be present wherever loss occurs.

The Red Cross does not make war; it responds to war's consequences. And in that response, it carries forward a tradition as old as Solferino. Humanity persists, even in conflict, and someone must stand in the gap between violence and care.

THE RETURN RITUAL: PUBLIC GRIEF AND PERSONAL SILENCE

"They told me to stand tall beside the casket. So I did. But inside, I was breaking."—Military escort, Afghanistan War

In modern war, death comes home with ceremony.

The rituals of return, the transfers, the escorts, the choreographed journey home, are precise and reverent. But they cannot account for everything that happens after.

For the escorts who accompany the fallen, the role demands emotional containment. It is a role that demands emotional containment.

Many escorts are young. Some knew the deceased; others did not. All of them bear the weight of proximity to grief without permission to express their own. "I couldn't cry," one soldier said. "Not in front of the family. Not on the plane. Not at the grave. So I didn't." But later, he admitted, he had broken down in a gas station bathroom—alone, anonymous, finally unable to carry the silence.

For the families, too, grief is both public and private. The thirteen folds have been performed for nearly a century now, the same careful geometry, the same transfer of weight from uniformed hands to grieving ones. Gold Star parents receive the flag, the formal condolences, the twenty-one-gun salute. But they also receive emails, comments, messages from strangers.They are expected to be strong, to be grateful, to endure. Some do. Others collapse under the weight of performance.

In the age of social media, mourning is often lived online. Tribute videos are made. Facebook pages become memorials. Memorial hashtags trend for a day, a week, then fade. For some families, these digital rituals offer connection. For others, they feel invasive, their grief consumed by a public that moves on too fast.

The military, for its part, has expanded grief support programs: casualty assistance officers, survivor outreach services, chaplaincy networks. And yet, many families still report feeling lost, adrift in a sea of bureaucracy and quiet expectation. They are told, implicitly, to grieve with grace—but not too loudly. To remember—but to move forward.

And what of those who handled the body?Who zipped the bag, cleaned the blood, loaded the aircraft?

Their grief has no ceremony.

There are no rituals for the medics who couldn't save them, the soldiers who found them, the contractors who carried the remains. They are left only with memory. And silence.

Modern return rituals do what they are designed to do. They offer dignity, precision, and national respect. But they cannot address the full scope of grief. They cannot hold the unspoken questions: *Why him? Why now? What was it for?*

These questions haunt the escorts, the mortuary teams, the families, and the nation.

Because when a war ends, its rituals linger. They become muscle memory. They become silence. They become stories waiting to be told.

When the System Failed: The Dover Scandal and the Cost of Neglect

"We trusted them with our son's body. And they threw him away." Mother of fallen Marine, 2011

The promise is sacred: that every fallen service member will be treated with dignity, identified with care, and returned to their family whole. For most of the modern era, the military has kept that promise with extraordinary dedication. But in 2011, the nation learned that the promise had been broken.

An investigation by the *Washington Post* revealed that the Port Mortuary at Dover Air Force Base—the facility that receives every American service member killed overseas—had disposed of partial remains in a landfill.[27] Not once. Not by accident. But systematically, over years, without the knowledge or consent of the families involved.

The remains were fragments—small portions that had been identified after initial burial, or additional pieces recovered from crash sites and battlefields. Military protocol required that such remains be treated with the same dignity as complete bodies, prepared, honored, and returned to families for reinterment. Instead, hundreds of portions were cremated and discarded with medical waste.

The scandal exposed failures at every level. Whistleblowers within the mortuary had raised concerns years earlier, only to be ignored or retaliated against.[28] Supervisors prioritized efficiency over reverence. Oversight mechanisms failed to detect systematic violations. The institution designed to honor the dead had, in some cases, dishonored them.

For families who learned that their loved ones' remains had been treated this way, the betrayal was devastating. They had trusted the military with their most sacred obligation. That trust had been violated in ways they could never have imagined.

The response was significant. The Air Force launched multiple investigations. Personnel were disciplined, and procedures were reformed. A new standard requiring family consent before any disposition of remains was implemented.[29] The Secretary of the Air Force apologized publicly, acknowledging failures that should never have occurred.

But the damage could not be undone.

The Dover scandal revealed something this book has emphasized throughout: that care for the dead is not automatic. It requires vigilance. It requires institutional commitment. It requires individuals who are willing to speak up when standards slip. And it requires a public that pays attention, understands the sacred nature of this work, and demands accountability when it fails.

The men and women who work at Dover today are acutely aware of this history. Many joined the mortuary system precisely because they believe in its mission. They carry the weight of the scandal not as personal guilt, but as institutional memory, a reminder of what happens when the system is taken for granted.

The scandal also reinforced that care must be guarded continually. The GRS soldiers of World War I and the Mortuary Affairs specialists of Vietnam worked within systems that could fail if vigilance lapsed; the same is true for the forensic teams of the DPAA. The dead cannot advocate for themselves. Their dignity depends entirely on the living.

Dover has recovered. The Port Mortuary today operates with standards that are among the most rigorous in the world. But the scandal remains a cautionary tale, a reminder that even the most sacred institutions can falter and that the promise to honor the dead must be renewed, actively, every day.

The fallen deserve nothing less.

Some Gave All: The Myth of a Grateful Nation

"They said, 'Thank you for your service.' But they didn't want to hear what the service did to us."—Iraq War veteran, Mortuary Affairs

We are told that ours is a grateful nation. That we honor our fallen. That we support our veterans. That we never forget.

And yet, for many who served—not just in combat, but in the shadows of deathcare—gratitude has felt more like performance than promise.

The phrase "Some Gave All" adorns bumper stickers, memorial walls, T-shirts, and campaign ads. It speaks to the ultimate sacrifice. But beneath the slogan lies a quieter truth. Some gave all, yes—but others gave everything they could and still returned to a country that did not know what to do with them.

Post-9/11 America built an identity around supporting the troops. Airports began applauding soldiers in uniform. Politicians ended speeches with tributes to the fallen. Entire nonprofits were founded to assist veterans and their families. But for those who served in mortuary affairs, or as combat medics, or contractors, or grief escorts, this support often rang hollow.

They were thanked for their service. But not asked what their service required.

Many returned home struggling with moral injury—not from killing, but from caring. From handling the dead. From delivering the bad news. From carrying caskets into gymnasiums, churches, and muddy hometown cemeteries. They carried grief, and often, they carried it alone.

Some sought help from the Veterans Affairs (VA) system, only to encounter long waitlists, red tape, or disbelief. Others never sought help at all, convinced that because they hadn't been "in combat," their trauma didn't count.

The truth is that trauma doesn't need a firefight.It only needs a memory that won't let go.

The suicide epidemic among post-9/11 veterans is a national s hame.Twenty-two a day.Many of them had never fired a weapon in anger—but had seen things they couldn't unsee.Body parts. Letters in pockets. Prayers written in sand.

Some were the ones who zipped the bags. Others were the ones who carried them.

And then there are the families—the Gold Stars and the grieving. They are given folded flags and final salutes. They are told their loss has meaning. But often, they are left to make meaning on their own.

In this country, we do grief by ceremony and slogan, but not always by action.We do tribute—but often not treatment.We honor the dead—but we neglect the living.

And so the question echoes: *what does a grateful nation really look like?*

It looks like remembering the names that never made headlines.It looks like funding mental health for the ones who carry death back home.It looks like hearing the stories that aren't easy to tell.And it looks like facing the uncomfortable truth that ritual without care is empty.

"Some gave all."Yes.But some gave everything they had to bring the fallen home.And they deserve more than a phrase.

They deserve to be seen. To be supported. To be remembered—not just for what they buried, but for what they bore.

THE FINAL WATCH: WHEN VETERANS DIE AT VA HOSPITALS

"No one should die alone. Especially not someone who served."—No Veteran Dies Alone volunteer, 2022

The care of war dead does not end when the war ends. It extends through the lives of veterans to the moment of their deaths, sometimes decades after the conflict that marked them.

Every year, thousands of veterans die in VA hospitals and nursing homes across the country. Many have families nearby. Many do not. Some are estranged from loved ones. Some have outlived everyone who knew them. Some have been homeless, incarcerated, or institutionalized for years. When they reach their final hours, they may be surrounded by medical staff who do not know their names, let alone their service.[30]

In response, the VA has developed the No Veteran Dies Alone (NVDA) program. Trained volunteers sit vigil with dying veterans, ensuring that someone is present at the end. They hold hands. They speak quietly. They read poetry or scripture or simply sit in silence.

They bear witness to the final moments of lives that were shaped by service.[31]

"I don't always know their story," one NVDA volunteer explained. "Sometimes they can't speak by the time I arrive. But I know they served. And I know they shouldn't be alone."

The program emerged from a simple recognition: that the nation owes its veterans more than medical care. It owes them presence, companionship, and the dignity of being seen at the end.

For veterans of Mortuary Affairs—those who spent their service handling the dead—there is a particular poignancy to this final chapter. They processed thousands of remains. They ensured that others were not left alone, not abandoned, not forgotten. And now, at the end of their own lives, they depend on the same commitment being extended to them.

Some veterans receive full military burial honors: the folded flag, the rifle salute, the playing of Taps. National cemeteries, including Arlington, offer interment for those who served. But for veterans who die without family, without resources, or without connection to military burial systems, the end can be quieter. They are offered a hospital bed, a volunteer's hand, a simple service at a VA cemetery where rows of identical headstones mark the final resting place of those who served.[32]

The Missing in America Project works to identify and inter the unclaimed cremated remains of veterans—ashes that sit in funeral homes, forgotten or abandoned, waiting for someone to claim them.[33] When remains are identified as belonging to a veteran, the organization arranges for burial with military honors. The dead are finally given the recognition their service earned.

As a Service to Armed Forces volunteer, I visit veterans at the VA hospital each month. I have seen the toll that war takes—not just the physical wounds, but the psychological scars that last a lifetime. I have handed flag stars to veterans who wept at the touch of worn fabric. I have listened to stories that had been held in silence for decades.

And I have thought about what it means to care for these men and women not just during their lives, but at the moment of their deaths.

The caretakers of the war dead were promised that they would not be forgotten. The NVDA program, the national cemetery system, the volunteers who sit with the dying—these are attempts to keep that promise.

It is not enough. It will never be enough.

But it is the work we must do.

Because the care of the dead does not end when the body is buried. It extends to the moment of death and to the lives that precede it.

Veterans deserve to be honored not just in parades and speeches, but in hospital rooms and hospice beds. They deserve to die knowing that someone is there. That their service mattered. That they are not alone.

This, too, is burial work.

This, too, is sacred.

And every December, the work of remembrance takes physical form.

Wreaths Across America coordinates volunteers to place wreaths on veterans' graves at Arlington National Cemetery and more than 3,500 locations across the country.[34] The mission began in 1992, when a Maine wreath company donated surplus wreaths to Arlington. It has grown into a national movement involving hundreds of thousands of volunteers who gather on a single December Saturday to honor the dead.

The ritual is simple but profound. Each volunteer takes a wreath, walks to a grave, and speaks the veteran's name aloud before placing the wreath. "Say their name" is the program's guiding principle, a recognition that the dead are not abstractions but individuals who served, who lived, who deserve to be remembered by name.[35]

American Red Cross volunteers participate in Wreaths Across America events across the country, continuing the organization's long tradition of connecting communities to those who served.[36] For many volunteers, placing a wreath is their first encounter with a military cemetery, their first experience of standing among rows of identical headstones and understanding, viscerally, the scale of sacrifice.

For Gold Star families, the wreaths are a sign that their loved ones have not been forgotten—that strangers will walk among the graves and speak their names, even decades after their deaths.

This, too, is burial work. Not the digging of graves, but the tending of them. Not the preparation of bodies, but the preservation of memory. The dead require ongoing care. And every December, in cemeteries across the nation, volunteers provide it.

Always Coming Home

No matter how far the battlefield, no matter how modern the war, the fallen still come home.

They come in metal caskets, in flagged convoys, in silence.They come by helicopter, by transport plane, by the hands of those who walk beside them.And with them come those who have carried the cost—not in wounds, but in memory.

Modern war is a war of precision and power, but it is still a war of people. And death, no matter how clean the technology, is still a messy, human thing.

Behind every dignified transfer, every folded flag, every digital profile in a government database, there is someone who carried the body. Someone who saw. Someone who cannot forget.

The tools have changed.But the task has not.

We still need those who will hold the dead with reverence, who will kneel beside wreckage, who will lift while others look away from, who will preserve the sacred in a world that so easily forgets what is holy. These men and women—soldiers, contractors, medics, morticians, handlers—are the last to touch the fallen. They are the ones who keep the final promise: *you will not be left behind.*

They do not do it for glory. They do not do it for recognition. They do it because someone must.

Because grief still matters.Because ritual still matters.Because even in modern war, memory is carried one body at a time.

And so the work continues.

Right now, as you hold this book, someone is doing this work.

At Dover Air Force Base, a dignified transfer team is preparing for an arrival. The aircraft will land in darkness. The casket will be carried in silence. No cameras will be present unless the family has consented. But the ritual will unfold as it has unfolded thousands of times: precise, reverent, sacred.[37]

In a laboratory in Hawaii, a forensic anthropologist is analyzing bone fragments recovered from a Pacific island. The remains have waited eighty years for this examination. The family—a great-niece who never knew her great-uncle—has provided a DNA sample. The comparison will take months. But the work is underway.[38]

In Ukraine, a volunteer recovery team is exhuming bodies from a mass grave near Izium. They photograph each face. They collect DNA samples. They wrap the remains in white cloth and mark the graves with numbers, hoping that someday the numbers will become names.[39]

In Syria, a woman is searching for her son. He was detained in 2013. She has heard nothing since. Every few months, she submits another inquiry through the Red Cross. Every few months, she receives the same response: no information available. But she keeps searching, because to stop would be to abandon him.[40]

In a quiet house in Ohio, a Gold Star mother is winding a watch that stopped at the moment of the blast that killed her son. She winds it once a year, on the anniversary. It is her ritual. It is how she keeps time with him.

In a VA hospital in Texas, a former Mortuary Affairs specialist is finally talking about what he saw in Iraq. It has taken fifteen years. The counselor listens. The words come slowly, but they come.

This is the work.

It is not finished.

It will never be finished.

Because as long as wars are fought, someone must carry the dead.

Someone must speak the names.

Someone must remember.

They are working right now.

While we sleep, while we scroll, while we forget.

They remember.

And because of them, the fallen come home.

Lest We Forget the Ones Who Remembered

"We are the witnesses. And we are the ones who remem-

ber the witnesses."—Postwar journal, anonymous GRS clerk

History remembers the warfighter—the flag. The charge. The victory. But who remembers the silence that follows? Who remembers the ones who returned to the battlefield after the guns fell silent, not to claim glory, but to collect the dead?

This book began with a question—not about how we fight, but how we remember.How we honor the fallen.How we carry the grief they leave behind.How we remember the ones who remembered.

From the ancient trenches of Bronze-Age Europe to the digital databases of Dover, from the whispered chants of Indigenous burial rites to the DNA-matched fragments of a lost Marine, from the women who rolled bandages in Civil War kitchens to the contractors who knelt beside black bags in Iraq, they were always there.

They were there when the body was broken.When the name was lost.When the battlefield was so charred no grave could hold its story.

They were there when death became bureaucracy.When grief became protocol.When memory became metadata.

And they still return—again and again—not to fight, but to finish the war. Because no war ends until the dead are laid to rest.

We forget this.

In textbooks, we trace battles. We count the fallen. We name the generals. But the hands that carried the stretchers, the ones that folded uniforms, lifted limbs, dug graves in mud and moonlight, are too often erased.

This book has tried to bring them back.

We have heard from gravediggers who carved crosses in shell-pocked soil, from chaplains who blessed both friend and enemy, from families who held a pouch of ashes and called it homecoming, from medics who closed a thousand eyes—and couldn't close their own. We have followed the body from sand to sky, from scaffold to shrine, from dog tag to DNA. We have walked beside the caretakers, the memory workers, the invisible saints of aftermath.

And now, we must remember them.

Let this be our vow:

We will not speak of war without speaking of the ones who buried it.We will not say "freedom isn't free" without counting the cost in shovels and silence.We will not erect another monument until we ask—who built the last one?And who *carried* the name etched into stone?

Because war is not just fought on battlefields; it is ended in morgues. In temporary graves. In kitchen tables covered in condolence letters. In the quiet of those who walk beside the dead and whisper, *"You are not forgotten."*

The stretcher-bearer.The body escort.The forensic anthropologi st.The Gold Star parent.The Indigenous matriarch repatriating sacred bones.The chaplain at the edge of the trench.The GRS clerk who wrote names until her hands bled.

They are the custodians of closure.

And they do not carry rifles.They carry reverence.

Their war is not for territory.It is for memory.

Their mission is not to destroy.It is to rebuild identity, dignity, and the threads of our shared humanity.

And so, when we say *"Lest we forget,"* let us mean it.

Let us remember not only the dead, but the ones who *laid them down*. Who bore them home.Who knelt in the mud while the rest of us looked away.

These are the ones *known unto us now*.

Because we have looked for them. Because we have said their names. Because we have remembered.

In doing so, we restore not just their stories but our own. Because to honor the dead, we must first honor the ones who did that work.

And so, one final time, let the record show:

They were there.

And now, we see them.

EPILOGUE

WHAT IT MEANS TO REMEMBER

"They are not buried in silence. We remember them out loud."—Flanders Field American Cemetery, Belgium

One book will not end war. Not mine, not anyone's. Not *All Quiet on the Western Front*, not *A Farewell to Arms*, not the photographs from Abu Ghraib, not the footage from Bucha. Humanity has been shown, again and again, what war does to bodies and souls—and humanity has continued to make war.

If documentation alone could end suffering, suffering would have ended long ago.

But.

Books do something else. Something slower. Something that matters even if it doesn't produce immediate, measurable peace.

Books create witnesses.

Every person who reads this book will carry something they didn't before. They will see the burial crews. They will think about the hands that fold the flags. They will pass a military cemetery and feel a new weight. They will watch the news differently. They will ask different questions.

I am not ending war; I am changing consciousness, one reader at a time. I am making the invisible visible. I am ensuring that the caretakers are no longer alone with what they carry.

That is not nothing.

That is everything I can do with words on a page.

***

I walked among the graves at Flanders and felt the breath leave my body.

The rows stretched endlessly—white crosses, perfectly spaced, each marking a name, a date, a life folded too soon. The air was still. The grass clipped to reverence. The only sound was my footsteps—until a single note, soft and solemn, pierced the quiet. "Taps." A lone trumpet echoing across the stones like a prayer held in brass.

I had come to Belgium following the trail of history. But what I found was something older, something holy: *memory made sacred.*

There, I learned that every American soldier buried at Flanders had been *adopted* by a Belgian family. They tend the graves. They place flowers. They stand vigil on Memorial Day—not out of duty, but out of love. Because they remember. Because *someone must.*

It broke something open in me.

As a Red Cross volunteer who has helped connect military families with their loved ones, who has taught the laws that protect the dead, I thought I understood this work intellectually. But standing there, I understood it in my body. This was not history. This was sacred ground.

In that stillness, I did not think of war as past. I thought of the ones who had *buried* it. The men who carried these bodies across blasted fields. The women who wrapped keepsakes in muslin and packed footlockers with shaking hands. The clerks who transcribed death into ledgers.The chaplains who whispered blessings to the wind. The medics who wept in silence.

They, too, deserve to be mourned. They, too, deserve to be remembered.

That silence—*their* silence—is what birthed this book.

***

I am not a soldier. I am a descendant. A witness. A gatherer. A keeper of the names that history forgot.

My family carries war in quiet ways. In the stories we don't tell, in the eyes of those who came home different, in the folded flags tucked in drawers, and the photographs yellowing in albums. We carry the ache of survival. And I carry that ache into my work, my writing, my life.

If you carry one thing from this book, let it be this:

War may end in silence. But memory begins with voice.

To speak the names of those who carried the dead is to carry them home at last.

Post Script

Questions for Remembrance and Relfection

This book has explored the quiet, sacred labor of those who buried the dead in war—men and women who stood between violence and memory and carried home more than bodies. These questions are offered not merely as discussion prompts, but as invitations to reflect, to reckon, and to remember.

1. What does it mean to remember the rememberers? Why is this act of remembrance essential in our time?

2. How did this book reshape or deepen your understanding of military service, sacrifice, and the unseen labor of care?

3. Many individuals in this story never fired a weapon but were forever changed by war. What forms of invisible trauma—moral, emotional, spiritual—did they carry?

4. How do cultural burial rituals—ancient or modern—help restore humanity after war? What role do ceremony and ritual play in healing and identity?

5. How does the historical marginalization of caregivers—especially women, people of color, Indigenous nations, and the poor—reveal the biases of how we record war?

6. Do you know someone in your own life who served in war but never spoke of what they saw? What might that silence carry?

7. In an age of fast news and digitized mourning, how do we preserve space for real grief—public and private?

8. What responsibility do we bear, as individuals and as a society, to those who served quietly and came home changed?

9. If you could speak to one of the forgotten caretakers of war, what would you want to say?

10. What legacy will you carry forward from this story—for your family, your community, and your descendants?

Why I Wrote This Book

I wrote this because silence isn't peace. Because forgetting doesn't heal. Because absence is not the same as closure.

I wrote this because I was tired of walking through history books and not seeing the *caretakers*. The women, the laborers, the marginalized, the ones who carried the wounded, buried the dead, and held memory together while the world collapsed.

I wrote this because war doesn't just belong to generals and timelines. It belongs to the ones who folded letters into bloodstained pockets. Who cooked for the dying, wept beside graves with no markers, and remembered.

I wrote this because I believe stories—*true* stories—can be medicine. Not just for the living but for the forgotten.

I am a witness. A gatherer. A keeper of names.

I don't write because I have all the answers. I write because I believe remembering is a form of resistance. And a form of love.

I wrote this because I believe in a better future. A world where peace is not an afterthought but a shared pursuit.

A world where our children protect oceans, not oil fields. Where justice grows out of memory, not denial. Where no one is ever asked to disappear quietly again.

***

This book is my offering.

To the silenced. To the forgotten. To the brave.

To the ones who buried the dead and were buried by history in return.

To the ones who remembered—so the rest of us might begin to.

***

For my grandchildren, curious, brilliant, and full of light, I want you to know:

You come from people who remembered.

You come from those who dug graves so others could rest.

You come from hands that bore the unbearable and still chose dignity.

Let this be your inheritance, not just the history, but the compassion.

Remember not only the fallen. Remember the ones who buried them.

And know that care, even in the face of horror, is resistance.

And to you, dear reader:

Thank you. For standing witness. For choosing light over silence.

May we never forget the ones who remembered.

What You Can Do

"The work of remembrance is not finished. It is handed forward."

You have read this book. You have walked beside the caretakers through trenches and temporary cemeteries, through mortuary tents and mass graves, through the silence of those who carried the dead and the longer silence of those who forgot them.

Now you carry something too.

Not the weight of bodies. Not the memory of faces. But the knowledge that this work exists, that it has always existed, and that it continues right now. While you hold this volume, someone is folding a flag, someone is excavating a crash site, someone is sitting with a dying veteran, someone is searching for the missing.

You cannot undo what the caretakers have borne. But you can ensure they are not alone in bearing it. You can become part of the continuum of care that stretches from Clara Barton to the volunteer who will place a wreath next December, from Henri Dunant at Solferino to the ICRC delegate documenting the dead in Syria today.

Here is how.

LISTEN

The most powerful thing you can do may also be the simplest: listen to those who have served.

If you know a veteran—a neighbor, a relative, a friend—consider asking them not about combat, but about what came after. What did they see? What did they carry? What has stayed with them?

Many will not answer. The silence is its own response, and it deserves respect. But some have been waiting decades for someone to ask. They have held stories that no one wanted to hear. They have carried weight that no one acknowledged.

Your willingness to listen without judgment, without trying to fix, and without turning away may be the first time anyone has offered to share their burden.

Listening is not passive; it is an act of witness. And witnesses are what the caretakers have always needed.

VOLUNTEER

The organizations that support military families, veterans, and the missing need hands and hearts. Consider offering yours.

American Red Cross Service to Armed Forces provides emergency communication, support to military families, and services to veterans in hospitals and transition. Volunteers visit VA hospitals, assist with casework, and help families navigate the particular challenges of military life and loss. No medical background is required—only the willingness to show up.

No Veteran Dies Alone (NVDA) trains volunteers to sit vigil with veterans in their final hours. Many veterans die in VA hospitals without family present. NVDA volunteers ensure that no one who served dies without someone beside them holding a hand, speaking quietly, or bearing witness to the end of a life shaped by service.

Wreaths Across America coordinates volunteers to place wreaths on veterans' graves each December at Arlington National Cemetery and over 3,500 locations nationwide. The ritual is simple: take a wreath, walk to a grave, speak the veteran's name aloud, and place the wreath. In that moment, you become a caretaker. You are tending a grave, honoring a life, refusing to let a name be forgotten.

Restoring Family Links, operated by the American Red Cross and the International Committee of the Red Cross, helps families separated by conflict, disaster, and displacement reconnect. Volunteers assist with casework, helping refugees and asylum seekers search for missing relatives. Clara Barton's work continues, and it needs you.

Women of the World and similar refugee resettlement organizations support forcibly displaced women and families as they rebuild their lives. Many of these families carry the weight of war—missing relatives, unburied dead, grief without closure. Your support helps them find stability and, sometimes, answers.

Missing in America Project identifies and inters the unclaimed cremated remains of veterans—ashes that sit in funeral homes, forgotten or abandoned. When remains are identified, the organization arranges burial with military honors. The dead are finally given the recognition their service earned.

Contact your local VA hospital, Red Cross chapter, or veterans' organization to learn what volunteer opportunities exist in your community. The need is constant. The work is waiting.

TEACH

The rules of war exist because people insisted on them. They are not natural laws but human choices, codified into treaties and conventions because previous generations understood what happens when war has no limits.

Teach your children that even in conflict, there are boundaries. Teach them that the wounded must be cared for, that prisoners must be protected, that the dead must be treated with dignity. Teach them that these are not suggestions but laws, and that someone must always be responsible for enforcing them.

The American Red Cross offers International Humanitarian Law education programs for youth and adults. The International Committee of the Red Cross provides educational resources on the Geneva Conventions and the laws of armed conflict. These materials are available online, in schools, and through community programs.

But you do not need a formal curriculum to teach. You can teach by example. You can teach by taking your children to a military cemetery and standing with them in silence. You can teach by telling them about the caretakers—the people who stay behind to do what no one else will do.

The next generation will inherit the world we leave them. They will face conflicts we cannot predict. What they carry into those conflicts—what values, what limits, what understanding of shared humanity—depends on what we teach them now.

The caretakers' burden is lightened when society understands what they carry. Education is prevention. And prevention is the highest form of care.

REMEMBER

Visit a military cemetery. Any cemetery will do—Arlington, yes, but also the small veterans' sections in local graveyards, the American cemeteries abroad, the markers in churchyards and town squares.

Walk the rows. Read the names. Notice the dates—how young so many of them were. Notice the wars—how many there have been. Notice the gaps—the unknown soldiers, the names worn away by weather, the ones who were never found.

Stand in the silence. Let yourself feel the weight of what you have never been asked to carry.

This is not morbid. It is not dwelling on death. It is the simple act of acknowledgment, recognition that these were real people. Someone buried them, someone mourned them, and they deserve to be remembered.

You might bring a flower. You might bring a flag. You might bring nothing but your presence. It is enough.

The caretakers have always known that the dead need witnesses. Now you know it too.

ADVOCATE

The systems that care for the dead and support those who carry them need resources, recognition, and reform. Your voice matters.

Support mental health services for veterans. Moral injury—the wound of participating in or witnessing events that violate one's moral code—is real, devastating, and often untreated. Advocate for expanded mental health resources, for the recognition of moral injury as a legitimate diagnosis, and for the destigmatization of seeking help.

Support the Defense POW/MIA Accounting Agency. The DPAA works to recover and identify the remains of more than 81,500 Americans still missing from past conflicts. Their work is painstaking, underfunded, and essential. Families are still waiting for answers. The promise made to the missing must be kept.

Support organizations that document the dead and missing in current conflicts. The International Committee of the Red Cross, Physicians for Human Rights, the International Commission on Missing Persons, and similar organizations work to identify the dead, support families, and hold perpetrators accountable. Their work is the continuation of everything this book has traced.

Support accountability. When atrocities occur, justice matters—not only for the victims but for the caretakers who process what atrocity leaves behind. International tribunals, truth commissions, and human rights documentation all serve to validate what witnesses have seen. When perpetrators are held accountable, the burden of witness is shared by society rather than borne by individuals alone.

Write to your representatives. Donate to organizations doing this work. Speak about what you have learned. The political will to care for the dead—and those who carry them—depends on a public that understands why it matters.

Create

If you are an artist, a writer, a filmmaker, a musician, consider how your gifts might serve this work.

The caretakers have been invisible in part because they have been absent from our stories. The war films focus on combat. The memorials honor the fallen. But the burial crews, the mortuary clerks, the families who waited—they appear only in the margins, if at all.

You can change that.

Write the poem about the gravedigger. Paint the image of the stretcher-bearer. Compose the song for the chaplain who blessed ten thousand bodies. Tell the story of the Gold Star mother, the DPAA anthropologist, the No Veteran Dies Alone volunteer.

Art makes the invisible visible. It creates empathy across distance and time. It ensures that what should not be forgotten cannot be forgotten.

This book is my contribution. What will yours be?

Begin

The work of care is not a single act. It is a practice. Something you return to, again and again, throughout your life.

You may volunteer once and then step back. You may listen to one story and carry it always. You may visit one cemetery and never forget what you felt there. That is enough. Every act of witness matters.

Or you may find, as I did, that this work becomes part of who you are. That once you see the caretakers, you cannot unsee them. That once you understand what they carry, you cannot pretend you do not know.

Either way, you are no longer the same person who opened this book.

You have become a witness.

And witnesses have obligations.

Not to fix everything or to end all suffering. Not to carry what is not yours to carry. But to see, to acknowledge, to refuse the comfort of forgetting. To ensure that the next generation knows what this generation learned.

The caretakers have waited a long time to be seen. Now they are seen by you.

What you do with that seeing is your choice.

But know this: every time you listen to a veteran, visit a grave, speak a name, teach a child, or simply pause to remember that somewhere, right now, someone is doing this sacred work, you join them.

You become part of the continuum of care.

And the burden they carry becomes, in some small measure, lighter.

Because they are no longer alone.

For information on volunteer opportunities:

- American Red Cross Service to Armed Forces: www.redcross.org/saf
- No Veteran Dies Alone: www.va.gov/geriatrics (search NVDA)
- Wreaths Across America: www.wreathsacrossamerica.org
- Restoring Family Links: www.familylinks.icrc.org
- Missing in America Project: www.miap.us
- Defense POW/MIA Accounting Agency: www.dpaa.mil
- International Committee of the Red Cross: www.icrc.org

For educational resources:

- American Red Cross IHL Programs: www.redcross.org/ihl
- ICRC Education Resources: www.icrc.org/en/war-and-law

APPENDIX

For Educators and Youth

A Guide for Using This Book in Educational Settings

Introduction for Educators

This book examines the care of war dead across history—a topic that intersects with history, civics, ethics, and international humanitarian law. The following materials are designed to help educators use this book with students at various levels, from middle school through college.

The themes of this book—death, trauma, grief, and war—require sensitive handling. Educators should preview content before assigning it and be prepared to support students who may have personal connections to military service, loss, or trauma. The goal is not to traumatize students but to help them understand the full scope of war's consequences and the humanity that persists even in conflict.

Discussion Questions by Level

Middle School (Grades 6–8)

1. What is a "caretaker of the dead?" Why is this job important?

2. Why do humans bury their dead? What does burial mean in different cultures?

3. Clara Barton founded the American Red Cross. What motivated her to help soldiers and their families?

4. What does it mean when we say that some people's contributions to history are "invisible?" Can you think of other examples of invisible workers?

5. Why is it important to identify soldiers who have died? How has technology changed the way we identify the dead?

6. What is the Tomb of the Unknown Soldier? Why do we honor someone whose name we don't know?

7. How do families cope when they don't know what happened to a loved one in war?

8. What is one thing you learned from this book that you didn't know before?

High School (Grades 9–12)

1. The author argues that "war does not end until the dead are buried." What does this mean? Do you agree?

2. How did the care of war dead change from the Civil War to

modern conflicts? What caused these changes?

3. What is "moral injury?" How is it different from PTSD? Why might someone who never fired a weapon experience moral injury?

4. How do different cultures approach death in war? What do these differences tell us about cultural values?

5. The book discusses how marginalized groups (women, Black soldiers, Indigenous peoples, immigrants) have contributed to burial work. Why have their contributions often been overlooked?

6. What responsibilities do governments have to identify and return the bodies of fallen soldiers? What about enemy soldiers?

7. How has technology changed grief? Is it easier or harder to mourn in the digital age?

8. What does it mean to be a "witness" to history? How does this book serve as an act of witness?

9. The author includes an apology to Vietnam veterans. Why does she feel this is necessary? Do you think public apologies can help heal historical wounds?

10. What would you want future generations to know about how we remember war?

College/Adult Education

1. How does the book challenge traditional military history

narratives? What voices are centered here that are typically absent?

2. Analyze the evolution of military mortuary practice as a reflection of changing attitudes toward death, bureaucracy, and national identity.

3. The book examines care as a form of "ethical resistance." What does this mean? Can care be a political act?

4. How do forensic anthropology and DNA technology intersect with ethics and cultural beliefs about the body?

5. The author positions this book as both historical scholarship and personal memoir. How does her family history and Red Cross work shape the narrative? Is this an effective approach?

6. Compare the treatment of burial rituals across cultures in Chapter 8. What universal themes emerge? What significant differences?

7. What obligations do democratic societies have to remember war's consequences, including the trauma borne by noncombat personnel?

8. How does the book's structure—interludes, spotlights, artifacts—contribute to its argument and emotional impact?

9. The Dover Air Force Base scandal revealed systemic failures in mortuary practice. What does this case study suggest about institutional accountability and the "guardianship" of sacred duties?

10. The book ends with ongoing work rather than resolution. Why is this significant? What responsibilities does this leave with the reader?

Connections to International Humanitarian Law

This book aligns with core principles of International Humanitarian Law (IHL), as codified in the Geneva Conventions. Educators teaching IHL can use this book to illustrate:

Protection of the Dead (Geneva Convention IV, Article 130; Additional Protocol I, Articles 33–34)

- The obligation to search for, identify, and bury the dead with respect
- The requirement to record burial locations and facilitate return of remains
- Prohibitions against despoliation and mistreatment of the dead

Discussion Questions for IHL:

1. What does IHL say about the treatment of the dead in armed conflict? Why are these rules necessary?
2. How does the principle of distinction (between combatants and civilians) apply to the care of the dead?
3. What happens when IHL rules about burial are violated? What are the legal, humanitarian, and psychological consequences?

4. The book discusses moments when enemies buried each other's dead. How does this relate to the IHL principle of common humanity?

5. What role does the International Committee of the Red Cross play in the care of the dead and missing?

ICRC YOUTH ACTION CAMPAIGN (YAC) ACTIVITIES

The following activities are designed for Red Cross Youth Action Campaign participants and other young humanitarians:

Activity 1: The Hidden Stories Project

Research a war memorial or military cemetery in your community. Find out:

- Who is buried or commemorated there?
- Who built the memorial or maintains the cemetery?
- Are there any "hidden" stories—contributions from women, minorities, or caregivers that are not prominently featured?

Create a presentation or social media campaign to share what you learn.

Activity 2: Interview a Witness

With appropriate preparation and adult support, interview a veteran, Gold Star family member, or humanitarian worker about their experiences with military death or grief. (Follow ethical guidelines for interviewing about sensitive topics.)

Questions to consider:

- What do you wish people understood about your experience?
- How has your experience shaped how you think about war and peace?
- What kind of support was helpful? What was missing?

Write a reflection or create a multimedia presentation about what you learned.

Activity 3: The Rules of War Simulation

Using the Geneva Conventions and this book as resources, create scenarios exploring ethical dilemmas related to the care of the dead:

- A soldier finds the body of an enemy combatant. What are their obligations under IHL?
- A humanitarian worker discovers a mass grave. What steps should they take?
- A family learns that their loved one's remains cannot be recovered. How should the military communicate this? What support should be offered?

Role-play the scenarios and discuss the ethical principles involved.

Activity 4: Restoring Family Links

Research the International Committee of the Red Cross's Restoring

Family Links program. Learn about:

- How the program works
- Historical examples of family tracing (from Clara Barton to today)
- Current conflicts where RFL is active

Create materials (posters, videos, social media content) to raise awareness about RFL in your community.

Activity 5: Memorial Design Challenge

Design a memorial for one of the "invisible" groups featured in this book:

- Civil War freedpeople who performed burial labor
- WWI Chinese Labour Corps
- WWII women clerks who typed death notifications
- Vietnam-era mortuary specialists
- Modern DPAA field teams

Consider: What would this memorial look like? Where would it be located? What message would it convey?

Present your design and explain your choices.

Activity 6: Letter to a Caretaker

Write a letter to someone featured in this book—a historical figure or a composite of the anonymous caretakers described. What would you want them to know? How would you honor their service?

Alternatively, write a letter to a modern caretaker: a DPAA anthropologist, a Dover Honor Guard member, a No Veteran Dies Alone volunteer.

Resources for Further Learning

Organizations:

- American Red Cross: www.redcross.org
- International Committee of the Red Cross: www.icrc.org
- Defense POW/MIA Accounting Agency: www.dpaa.mil
- Commonwealth War Graves Commission: www.cwgc.org
- No Veteran Dies Alone Program: www.va.gov/GERIATRICS/

Museums and Memorials:

- National WWI Museum (Kansas City, MO)
- National WWII Museum (New Orleans, LA)
- Vietnam Veterans Memorial (Washington, D.C.)
- Arlington National Cemetery
- Flanders Field American Cemetery (Belgium)
- In Flanders Fields Museum (Ypres, Belgium)

Primary Source Collections:

- Library of Congress Veterans History Project

- National Archives and Records Administration
- American Red Cross Archives

This educational appendix was developed by the author as a volunteer with the American Red Cross. For additional resources or educator support, contact the author, Sara Madden at sara.madden@redcross.org *or* contact@saramadden.com

Endnotes

Chapter One

1. Erella Hovers, et al., "An Early Case of Color Symbolism: Ochre Use by Modern Humans in Qafzeh Cave," *Current Anthropology* 44, no. 4 (2003): 491–522.
2. William Rendu, et al., "Evidence Supporting an Intentional Neandertal Burial at La Chapelle-aux-Saints," *Proceedings of the National Academy of Sciences* 111, no. 1 (2014): 81–86.
3. Ralph Solecki, *Shanidar: The First Flower People* (Knopf, 1971).
4. Mike Parker Pearson, *The Archaeology of Death and Burial* (Texas A&M University Press, 1999).
5. Kathleen Kenyon, *Digging Up Jericho* (Ernest Benn, 1957).
6. Ian Hodder, *Catalhoyuk: The Leopard's Tale* (Thames & Hudson, 2010).
7. James Brown, "The Search for Rank in Prehistoric Burials," *The Archaeology of Death,* edited by R. Chapman, I. Kinnes, and K. Randsborg (Cambridge University Press, 1981).
8. Euripides, *The Suppliants,* verse 670.
9. Ian Morris, *Death-Ritual and Social Structure in Classical Antiquity* (Cambridge University Press, 1992).
10. Sarah Tarlow, *Ritual, Belief and the Dead in Early Modern Britain and Ireland* (Cambridge University Press, 2011).
11. Jan Assmann, *Death and Salvation in Ancient Egypt,* trans. David Lorton (Cornell University Press, 2005), 23–31.
12. Salima Ikram, *Death and Burial in Ancient Egypt* (Longman, 2003), 48–52.
13. Christina Riggs, *Unwrapping Ancient Egypt* (Bloomsbury Academic, 2014), 67–72.

14. John H. Taylor, *Death and the Afterlife in Ancient Egypt* (University of Chicago Press, 2001), 112–118.

15. James Legge, trans., *The Li Ki (Book of Rites),* in *Sacred Books of the East,* Vol. 27 (Clarendon Press, 1885), 369–388.

16. Constance A. Cook, *Death in Ancient China: The Tale of One Man's Journey* (Brill, 2006), 89–95.

17. Robert Eno, *The Confucian Creation of Heaven* (SUNY Press, 1990), 134–142.

18. Namihira Emiko, "Pollution in the Folk Belief System," *Current Anthropology* 28, no. 4 (1987): 65–74.

19. Eiko Ikegami, *The Taming of the Samurai: Honorific Individualism and the Making of Modern Japan* (Harvard University Press, 1995), 287–294.

20. Stephen Turnbull, *The Samurai Sourcebook* (Cassell, 1998), 178–182.

21. Jonathan Parry, *Death in Banaras* (Cambridge University Press, 1994), 151–168.

22. J.A.B. van Buitenen, trans., *The Mahabharata,* Vol. 2 (University of Chicago Press, 1975), 418–426.

23. Wendy Doniger, trans., *The Rig Veda* (Penguin Classics, 1981), 49–54 (on cremation hymns).

24. Jack Goody, *Death, Property and the Ancestors: A Study of the Mortuary Customs of the LoDagaa of West Africa* (Stanford University Press, 1962), 34–48.

25. Kwame Gyekye, *An Essay on African Philosophical Thought: The Akan Conceptual Scheme* (Temple University Press, 1995), 156–164.

26. Robert S. Rattray, *Religion and Art in Ashanti* (Oxford University Press, 1927), 163–178.

27. Margaret Thompson Drewal, *Yoruba Ritual: Performers, Play, Agency* (Indiana University Press, 1992), 89–102.

28. Neil Price, *The Viking Way: Magic and Mind in Late Iron Age Scandinavia* (Oxbow Books, 2019), 77–85.

29. Karin Sanders, *Bodies in the Bog and the Archaeological Imagination* (University of Chicago Press, 2009), 112–118.

30. Neil Price, "Dying and the Dead: Viking Age Mortuary Behavior," in *The Viking World*, ed. Stefan Brink and Neil Price (Routledge, 2008), 257–273.

31. Martin Carver, *Sutton Hoo: Burial Ground of Kings?* (University of Pennsylvania Press, 1998), 165–189.

32. Price, *The Viking Way*, 301–312.

33. Howard Williams, *Death and Memory in Early Medieval Britain* (Cambridge University Press, 2006), 141–159.

34. Inga Clendinnen, *Aztecs: An Interpretation* (Cambridge University Press, 1991), 170–182.

35. Kay Almere Read, *Time and Sacrifice in the Aztec Cosmos* (Indiana University Press, 1998), 124–138.

36. Sabine MacCormack, *Religion in the Andes: Vision and Imagination in Early Colonial Peru* (Princeton University Press, 1991), 96–112.

37. Tom Dillehay, ed., *Tombs for the Living: Andean Mortuary Practices* (Dumbarton Oaks, 1995), 45–62.

38. Robert Garland, *The Greek Way of Death* (Cornell University Press, 1985).

39. Homer, *The Iliad*, trans. Robert Fagles (Penguin, 1990), Book 24.

40. Donna C. Kurts and John Boardman, *Greek Burial Customs* (Cornell University Press, 1971).

41. Thucydides, *History of the Peloponnesian War*, 2.34-2.46 (Pericles' Funeral Oration).

42. Karen Radner, "Assyria and the 3rd Millenium: The Historical Context," *State Archives of Assyria Bulletin* 12 (1998): 87–98.

43. John Curtis, *Ancient Persia: 550-330 B.C.* (British Museum Press, 1990).

44. Xuan Zhou, *The Book of Rites (Liji)*, Translated and annotated by James Legge (Dover, 2008).

45. Robert Eno, *The Confucian Creation of Heaven* (SUNY Press, 1990).

46. Daniel K. Richter, *Facing East from Indian Country: A Native History of Early America* (Harvard University Press, 2001).

47. Valerie Hope, *Roman Death: The Dying and the Dead in Ancient Rome* (Continuum, 2009).

48. Eve D'Ambra, *Roman Art* (Cambridge University Press, 1998).

49. Simon Hornblower, *The Greek World, 479-323 BC* (Routledge, 2002).

50. Ian Knight, *Zulu Rising: The Epic Story of Siandlwana and Rorke's Drift* (Thomas Dunne Books, 2010).

51. Jesse Byock, *The Saga of the Volsungs* (University of California Press, 1990).

52. John Bodel, "From Columbaria to Catacombs," in *Death and Disease in the Ancient City*, ed. Valerie Hope (London: Routledge, 2000), 105–127.

53. Éric Rebillard, *The Care of the Dead in Late Antiquity* (Cornell University Press, 2009), 32–40; 56-61.

54. Thomas Laqueur, *The Work of the Dead* (Princeton University Press, 2015), 68–72.

55. Valerie Hope, *Roman Death: The Dying and the Dead in Ancient Rome* (Continuum, 2009), 84–90.

56. Jörg Rüpke, *Religion of the Romans* (Polity Press, 2007), 214–220.

57. Ian Morris, *Death-Ritual and Social Structure in Classical Antiquity* (Cambridge University Press, 1992), 47–59.

58. Robin Jensen, *Understanding Early Christian Art* (Routledge, 2000), 57–65.

59. Rebillard, *Care of the Dead*, 91–97.

60. Ralph Solecki, "Shanidar IV, a Neanderthal Flower Burial in Northern Iraq," *Science* 190, no. 4217 (1975): 880–881.

61. Salima Ikram, *Death and Burial in Ancient Egypt* (Longman, 2003), 52–58.

62. Ann Rosalie David, *Handbook to Life in Ancient Egypt* (Oxford University Press, 1998), 335–342.

63. Christina Riggs, *Unwrapping Ancient Egypt* (Bloomsbury Academic, 2014), 74–79.

64. John H. Taylor, *Death and the Afterlife in Ancient Egypt* (University of Chicago Press, 2001), 186–191.

65. Edward F. Wente, trans., *Letters from Ancient Egypt* (Scholars Press, 1990), 142–145.

66. Zhongshu Wang, *Han Civilization,* trans. K.C. Chang (Yale University Press, 1982), 187–194.

67. Jessica Rawson, *Ancient China: Art and Archaeology* (British Museum, 1980), 142–148.

68. Robert Bagley, ed., *Ancient Sichuan: Treasures from a Lost Civilization* (Seattle Art Museum, 2001), 78–84.

69. Elizabeth Childs-Johnson, "The Jade Suit of Life," *Orientations* 27, no. 3 (1996): 54–61.

70. Constance A. Cook, *Death in Ancient China: The Tale of One Man's Journey* (Brill, 2006), 156–164.

71. Zhongshu Wang, *Han Civilization,* trans. K.C. Chang (Yale University Press, 1982), 194–198.

Chapter Two

1. Sophocles, *Antigone,* trans. Robert Fagles (Penguin, 1984), line 523.

2. Maurice Keen, *Chivalry* (Yale University Press, 1984).

3. Hugo Grotius, *On the Law of War and Peace,* ed. Stephen Neff (Cambridge University Press, 2012).

4. Maurice Keen, *Chivalry* (Yale University Press, 1984).

5. Christopher Tyerman, *God's War: A New History of the Crusades* (Harvard University Press, 2006).

6. Malcolm Vale, *War and Chivalry: Warfare and Aristocratic Culture in England, France, and Burgundy at the End of the Middle Ages* (University of Georgia Press, 1981).

7. Thomas Malory, *Le Morte d'Arthur,* Ed. Stephen H. A. Shepherd (Norton, 2004).

8. Jean Froissart, *Chronicles,* trans. Geoffrey Brereton (Penguin, 1968).

9. Alfred H. Burne, *The Crecy War: A Military History of the Hundred Years War from 1337 to the Peace of Bretigny, 1360* (Eyre & Spottiswoode, 1955).

10. Paul Binski, *Medieval Death: Ritual and Representation* (British Museum Press, 1996).

11. Caroline Walker Bynum, *The Resurrection of the Body in Western Christianity, 200-1336* (Columbia University Press, 1995).

12. Catechism of the Catholic Church, §2447.

13. Babylonian Talmud, Tractate Gittin 61a.

14. Mona Siddiqui, *How to Read the Qur'an* (Granta, 2007).

15. Mohammad H. Khalil, *Jihad, Radicalism, and the New Atheism* (Cambridge University Press, 2017).

16. Jonathan Parry, *Death in Banaras* (Cambridge University Press, 1994).

17. Raymond J. DeMallie, *The Sixth Grandfather: Black Elk's Teachings Given to John G. Neihardt* (University of Nebraska Press, 1984).

18. Catherine J. Allen, *The Hold Life Has: Coca and Cultural Identity in an Andean Community* (Smithsonian Institution Press, 2002).

19. John D. Kelly, *The American War in Contemporary Vietnam* (Cornell University Press, 2002).

20. U.S. War Department, *General Orders No. 100 (Lieber Code)*, 1863, Article 75.

21. Henri Dunant, *A Memory of Solferino* (International Committee of the Red Cross, 1862; repr. 1986), 40–52.

22. Henri Dunant, *A Memory of Solferino* (International Committee of the Red Cross, 1862; repr. 1986), 72.

23. Henri Dunant, *A Memory of Solferino* (International Committee of the Red Cross, 1862; repr. 1986), 115–128.

24. David P. Forsythe, *The Humanitarians: The International Committee of the Red Cross* (Cambridge University Press, 2005), 13–18.

25. Convention for the Amelioration of the Condition of the Wounded in Armies in the Field, August 22, 1864, 22 Stat. 940.

26. François Bugnion, *The International Committee of the Red Cross and the Protection of War Victims* (ICRC, 2003), 274–281.

27. Geoffrey Best, *Humanity in Warfare: The Modern History of the International Law of Armed Conflicts* (Columbia University Press, 1980), 128–135.

28. Caroline Moorehead, *Dunant's Dream: War, Switzerland, and the History of the Red Cross* (HarperCollins, 1998), 421–430.

29. Jost Dülffer, "The Hague Peace Conferences and the Limitation of Violence," in *The Cambridge History of War*, vol. 4, ed. Roger Chickering et al. (Cambridge University Press, 2012), 292–296.

30. James Brown Scott, ed., *The Hague Conventions and Declarations of 1899 and 1907* (Oxford University Press, 1915), 41–55.

31. Convention (II) with Respect to the Laws and Customs of War on Land and its annex: Regulations concerning the Laws and Customs of War on Land, July 29, 1899, 32 Stat. 1803.

32. Hague Regulations of 1899, Articles 1–20.

33. Hague Regulations of 1899, Article 14.

34. Convention (III) for the Adaptation to Maritime Warfare of the Principles of the Geneva Convention of 22 August 1864, July 29, 1899.

35. Geoffrey Best, *Humanity in Warfare: The Modern History of the International Law of Armed Conflicts* (Columbia University Press, 1980), 166–180.

36. Convention (IV) respecting the Laws and Customs of War on Land and its annex: Regulations concerning the Laws and Customs of War on Land, October 18, 1907, Article 19.

37. Hague Convention (IV) of 1907, Preamble.

38. Antony Anghie, *Imperialism, Sovereignty, and the Making of International Law* (Cambridge University Press, 2005), 115–140.

39. Geoffrey Best, *War and Law Since 1945* (Clarendon Press, 1994), 35–42.

40. Stanley Weintraub, *Silent Night: The Story of the World War I Christmas Truce* (Plume, 2002).

41. Malcolm Brown, *The Imperial War Museum Book of the First World War* (Pan Macmillan, 1993).

42. Tony Ashworth, *Trench Warfare 1914-1918: The Live and Let Live System* (Macmillan, 2000).

43. Stéphane Audoin-Rouzeau and Annette Becker, *14-18: Understanding the Great War* (Hill and Wang, 2002).

44. Jay Winter, *Sites of Memory, Sites of Mourning: The Great War in European Cultural History* (Cambridge University Press, 1995).

45. John Keegan, *The First World War* (Vintage Books, 1999), 422–428.

46. François Bugnion, *The International Committee of the Red Cross and the Protection of War Victims* (ICRC, 2003), 178–195.

47. Geneva Convention relative to the Treatment of Prisoners of War, July 27, 1929, 47 Stat. 2021.

48. Geneva Convention of 1929 (POW), Articles 1–8.

49. Geneva Convention of 1929 (POW), Article 76.

50. Geneva Convention of 1929 (POW), Article 77.

51. Geneva Convention of 1929 (POW), Articles 79, 86–88.

52. Bugnion, *The International Committee of the Red Cross*, 312–340.

53. Omer Bartov, *Hitler's Army: Soldiers, Nazis, and War in the Third Reich* (Oxford University Press, 1991), 83–105.

54. Geoffrey Best, *War and Law Since 1945* (Clarendon Press, 1994), 80–95.

55. Michael Sledge, *Soldier Dead: How We Recover, Identify, Bury, and Honor Our Military Fallen* (Columbia University Press, 2005), 98–125.

56. Jean S. Pictet, ed., *Commentary on the Geneva Convention for the Amelioration of the Condition of the Wounded and Sick in Armed Forces in the Field* (ICRC, 1952), 5–12.

57. Geneva Conventions of August 12, 1949 (I–IV).

58. Geneva Convention (I) of 1949, Article 17.

59. Geneva Convention (I) of 1949, Article 17.

60. Geneva Convention (III) of 1949, Article 120.

61. Geneva Convention (IV) of 1949, Articles 129–131.

62. Common Article 3, Geneva Conventions of 1949.

63. Protocol Additional to the Geneva Conventions of 12 August 1949, and relating to the Protection of Victims of International Armed Conflicts (Protocol I), June 8, 1977; Protocol Additional to the Geneva Conventions of 12 August 1949, and relating to the Protection of Victims of Non-International Armed Conflicts (Protocol II), June 8, 1977.

64. Protocol I of 1977, Article 33.

65. Protocol I of 1977, Article 34.

66. International Committee of the Red Cross, "State Parties to the Geneva Conventions," accessed 2024, https://www.icrc.org.

67. ICRC, *Commentary on the First Geneva Convention: Convention (I) for the Amelioration of the Condition of the Wounded and Sick in Armed Forces in the Field*, 2nd ed. (Cambridge University Press, 2016).

68. ICRC, *International Humanitarian Law and the Challenges of Contemporary Armed Conflicts* (ICRC, 2019), 7–22.

69. François Bugnion, *The International Committee of the Red Cross and the Protection of War Victims* (ICRC, 2003), 89–95.

70. Gradimir Djurovic, *The Central Tracing Agency of the International Committee of the Red Cross* (Henry Dunant Institute, 1986), 34–48.

71. Djurovic, *The Central Tracing Agency*, 52–67.

72. Bugnion, *The International Committee of the Red Cross*, 312–340.

73. Djurovic, *The Central Tracing Agency*, 89.

74. Jean-Claude Favez, *The Red Cross and the Holocaust* (Cambridge University Press, 1999), 267–285.

75. Bugnion, *The International Committee of the Red Cross*, 589–612.

76. Djurovic, *The Central Tracing Agency*, 156–162.

77. ICRC, *Restoring Family Links: A Guide for National Red Cross and Red Crescent Societies* (ICRC, 2001), 23–28.

78. ICRC, "Restoring Family Links Strategy 2020-2025" (ICRC, 2020), 4–8.

79. ICRC, *Annual Report 2022* (ICRC, 2023), 47–52.

80. ICRC, *Restoring Family Links: A Guide for National Red Cross and Red Crescent Societies*, 45–56.

81. ICRC, *Restoring Family Links: A Guide for National Red Cross and Red Crescent Societies*, 67–78.

82. ICRC, "Certificates and Attestations," accessed 2024, https://www.icrc.org.

83. Mark Skinner, Djordje Alempijevic, and Marija Djuric-Srejic, "Guidelines for International Forensic Bio-archaeology Monitors of Mass Grave Exhumations," *Forensic Science International* 134, no. 2–3 (2003): 81–92.

84. ICRC, *The Missing and Their Families: Summary of the Conclusions Arising from Events Held Prior to the International Conference of Governmental and Non-Governmental Experts* (ICRC, 2003), 34-42.

85. International Commission on Missing Persons, "History," accessed 2024, https://www.icmp.int.

86. ICMP, *Annual Report 2023* (ICMP, 2024), 12.

87. ICRC, *Guiding Principles/Model Law on the Missing* (ICRC, 2009), 8–15.

88. ICRC, *The Missing: ICRC Action to Resolve the Problem of People Unaccounted for as a Result of Armed Conflict or Internal Violence—A Framework for Reference* (ICRC, 2003).

89. ICRC, *Missing People, DNA Analysis and Identification of Human Remains: A Guide to Best Practice in Armed Conflicts and Other Situations of Armed Violence*, 2nd ed. (ICRC, 2009).

90. Defense POW/MIA Accounting Agency, "About DPAA," accessed 2024, https://www.dpaa.mil.

91. DPAA, "Our Missing," accessed 2024, https://www.dpaa.mil.

92. ICRC, *Restoring Family Links Strategy 2020-2025*, 28–32.

93. American Red Cross, "Holocaust and War Victims Tracing," accessed 2024, https://www.redcross.org.

94. ICRC, "International Conference of the Red Cross and Red Crescent," accessed 2024, https://www.icrc.org.

95. Pauline Boss, *Ambiguous Loss: Learning to Live with Unresolved Grief* (Harvard University Press, 1999), 3–18.

96. ICRC, *Accompanying the Families of Missing Persons: A Practical Handbook* (ICRC, 2013), 45–58.

97. ICRC, *Accompanying the Families of Missing Persons*, 78–92.

98. Alexander Laban Hinton, *Why Did They Kill?: Cambodia in the Shadow of Genocide* (University of California Press, 2005).

99. Martin Dean, *Collaboration in the Holocaust: Crimes of the Local Police in Belorussia and Ukraine, 1941-44* (Palgrave Macmillan, 2000).

100. David Chandler, *Voices from S-21: Terror and History in Pol Pot's Secret Prison* (University of California Press, 1999).

101. Mahmood Mamdani, *When Victims Become Killers: Colonialism, Nativism, and the Genocide in Rwanda* (Princeton University Press, 2001).

102. Human Rights Watch, "Between a Drone and Al-Qaeda: The Civilian Cost of U.S. Targeted Killings in Yemen," accessed April 18, 2025, https://www.hrw.org.

103. Eric Stover and Rachel Shigekane, "The Missing in the Aftermath of War: When Do the Needs of the Victims' Families and International War Crimes Tribunals Clash?," *International Review of the Red Cross* 84, no. 848 (2002): 845–865.

104. Francisco Ferrándiz and Antonius C.G.M. Robben, eds., *Necropolitics: Mass Graves and Exhumations in the Age of Human Rights* (University of Pennsylvania Press, 2015).

105. Peter Metcalf and Richard Huntington, *Celebrations of Death: The Anthropology of Mortuary Ritual* (Cambridge University Press, 1991).

106. Cheryl J. LaRoche and Michael L. Blakey, "Seizing Intellectual Power: The Dialogue at the New York African Burial Ground," *Historical Archaeology* 33, no. 3 (1999): 84–106.

107. Truth and Reconciliation Commission of Canada, *Final Report*, Vol. 1 (Montreal: McGill-Queen's University Press, 2015).

108. Judith Butler, *Precarious Life: The Powers of Mourning and Violence* (Verso, 2004).

109. Commonwealth War Graves Commission, *Special Committee Report on Historical Inequalities in Commemoration* (CWGC, 2021).

110. John David Smith, *Black Soldiers in Blue: African American Troops in the Civil War Era* (University of North Carolina Press, 2002).

111. Robert Asahina, *Just Americans: How Japanese Americans Won a War at Home and Abroad* (Gotham Books, 2006).

112. John W. Dower, *War Without Mercy: Race and Power in the Pacific War* (Pantheon, 1986).

113. Human Rights Watch, *World Report 2023.*

114. Mattie E. Treadwell, *The Women's Army Corps* (U.S. Army Center of Military History, 1954).

115. Arlington National Cemetery, "Eligibility and Burial Honors."

116. Antony Beevor, *The Second World War* (Little, Brown and Company, 2012).

Chapter Three

1. Drew Gilpin Faust, *This Republic of Suffering: Death and the American Civil War* (Vintage, 2008).
2. U.S. War Department, Series I, VO. 19 (Government Printing Office, 1887).
3. Caroline E. Janney, *Remembering the Civil War: Reunion and the Limits of Reconciliation* (Harvard University Press, 2001).
4. David W. Blight, *Race and Reunion: The Civil War in American Memory* (Harvard University Press, 2001).
5. Gary Laderman, *The Sacred Remains: American Attitudes Toward Death* (Yale University Press, 1996).
6. Michael J. Bennett, *Union Jacks: Yankee Sailors in the Civil War* (University of North Carolina Press, 2004), 145–62.
7. Craig L. Symonds, *The Civil War at Sea* (Oxford University Press, 2012), 78–92.
8. U.S. Navy, *Regulations for the Government of the United States Navy* (Government Printing Office, 1865), 112–14.
9. William Marvel, *The Monitor Chronicles: One Sailor's Account* (Simon & Schuster, 2000), 189–205.

10. NOAA, "USS Monitor: Recovery and Conservation," accessed 2024, https://monitor.noaa.gov/.

11. John R. Neff, *Honoring the Civil War Dead: Commemoration and the Problem of Reconciliation* (University Press of Kansas, 2005).

12. Drew Gilpin Faust, *This Republic of Suffering: Death and the American Civil War* (Vintage, 2008).

13. Kurt G. Piehler, *Remembering War the American War* (Smithsonian Institution Press, 1995).

14. Caroline E. Janney, *Burying the Dead but Not the Past: Ladies' Memorial Associations and the Lost Cause* (University of North Carolina Press, 2008).

15. David John Smith, *Black Soldiers in Blue: African American Troops in the Civil War Era* (University of North Carolina Press, 2002).

16. David W. Blight, *Race and Reunion: The Civil War in American Memory* (Harvard University Press, 2001).

17. Arlington National Cemetery, "History of Arlington," https://www.arlingtoncemetery.mil/Explore/History

18. U.S. War Department, *The War of the Rebellion: Official Records of the Union and Confederate Armies, Series II,* Vol. 5 (Government Printing Office, 1897).

19. G. Kurt Piehler, *Remembering War the American Way* (Smithsonian Institution Press, 1995).

20. Drew Gilpin Faust, *This Republic of Suffering: Death and the American Civil War* (Vintage, 2008).

21. William Corby, *Chaplain Life: Three Years in the Irish Brigade with the Army of the Potomac* (University of Notre Dame Press, 1992).

22. Richard Hall, *Women on the Civil War Battlefront* (University Press of Kansas, 2006).

23. Caroline E. Janney, *Burying the Dead but Not the Past: Ladies' Memorial Associations and the Lost Cause* (University of North Carolina Press, 2008).

24. National Park Service, "Clara Barton and the Office of Missing Soldiers," https://www.nps.gov/people/clara-barton-and-the-office-of-missing-soldiers.htm

25. Drew Gilpin Faust, *This Republic of Suffering: Death and the American Civil War* (Alfred A. Knopf, 2008), 99–103.

26. Jessica Mitford, *The American Way of Death Revisited* (Alfred A. Knopf, 1998), 44–50.

27. Gary Laderman, *The Sacred Remains: American Attitudes Toward Death, 1799–1883* (Yale University Press, 1996), 22–28.

28. Faust, *This Republic of Suffering*, 103–110.

29. Laderman, *The Sacred Remains*, 124.

30. Pat Jalland, *Death in the Victorian Family* (Oxford University Press, 1996), 147.

31. Joelle Gill, *The Modern Funeral Industry and the Civil War* (University Press of Mississippi, 2005), 77–79.

32. Mitford, *American Way of Death Revisited*, 52–57.

33. Jalland, *Death in the Victorian Family*, 148.

34. Laderman, *The Sacred Remains*, 128–130.

35. Jason Emerson, *Lincoln the Inventor* (Southern Illinois University Press, 2014), 201–210.

36. Faust, *This Republic of Suffering*, 140–141.

37. Gill, *Modern Funeral Industry*, 89.

38. Jeff Rosenheim, *Photography and the American Civil War* (Metropolitan Museum of Art/Knopf, 2013), 65–68.

39. Susan Sontag, *Regarding the Pain of Others* (Farrar, Straus and Giroux, 2003), 47–50.

40. *New York Times*, "Brady's Photographs of the Dead," October 20, 1862.

41. Rosenheim, *Photography and the American Civil War*, 12.

42. Drew Gilpin Faust, *This Republic of Suffering: Death and the American Civil War* (Vintage, 2008).

43. Richard Hall, *Women on the Civil War Battlefront* (University Press of Kansas, 2006).

44. Catherine Clinton and Nina Silber, eds., *Divided Houses: Gender and the Civil War* (Oxford University Press, 1992).

45. National Park Service, "Clara Barton and the Office of Missing Soldiers," https://www.nps.gov/people/clara-barton-and-the-office-of-missing-soldiers.htm

46. Caroline E. Janney, *Burying the Dead but Not the Past: Ladies' Memorial Associations and the Lost Cause* (University of North Carolina Press, 2008).

47. Karen L. Cox, *Dixie's Daughters: The United Daughters of the Confederacy and the Preservation of Confederate Culture* (University Press of Florida, 2003).

48. David W. Blight, *Race and Reunion: The Civil War in American Memory* (Harvard University Press, 2001).

49. Daniel Bellware and Richard Gardiner, *The Genesis of the Memorial Day Holiday in America* (Columbus State University, 2014).

50. Stephanie McCurry, *Confederate Reckoning: Power and Politics in the Civil War South* (Harvard University Press, 2010), 263–278.

51. Amy Murrell Taylor, *Embattled Freedom: Journeys through the Civil War's Slave Refugee Camps* (University of North Carolina Press, 2018), 45–72.

52. Ira Berlin et al., eds., *Freedom: A Documentary History of Emancipation, 1861-1867*, Series I, Vol. 1: *The Destruction of Slavery* (Cambridge University Press, 1985), 678–692.

53. Stephen W. Sears, *Landscape Turned Red: The Battle of Antietam* (Houghton Mifflin, 1983), 312–315.

54. Timothy B. Smith, *The Untold Story of Shiloh: The Battle and the Battlefield* (University of Tennessee Press, 2006), 134–142.

55. Drew Gilpin Faust, *This Republic of Suffering: Death and the American Civil War* (Vintage, 2008), 66-71.

56. Mechal Sobel, *The World They Made Together: Black and White Values in Eighteenth-Century Virginia* (Princeton University Press, 1987), 214–229.

57. Albert J. Raboteau, *Slave Religion: The "Invisible Institution" in the Antebellum South* (Oxford University Press, 2004), 243–266.

58. Works Progress Administration, Federal Writers' Project, *Slave Narratives: A Folk History of Slavery in the United States*, Vol. 14: South Carolina Narratives (Library of Congress, 1941), 287.

59. Jim Downs, *Sick from Freedom: African-American Illness and Suffering during the Civil War and Reconstruction* (Oxford University Press, 2012), 23–48.

60. Joseph T. Glatthaar, *Forged in Battle: The Civil War Alliance of Black Soldiers and White Officers* (Free Press, 1990), 169–183.

61. Glatthaar, *Forged in Battle*, 201–218.

62. John Cimprich, *Fort Pillow, a Civil War Massacre, and Public Memory* (Louisiana State University Press, 2005), 82–99.

63. John David Smith, ed., *Black Soldiers in Blue: African American Troops in the Civil War Era* (University of North Carolina Press, 2002), 1–28.

64. Joseph T. Glatthaar, *Forged in Battle: The Civil War Alliance of Black Soldiers and White Officers* (Free Press, 1990), 169–183.

65. Glatthaar, *Forged in Battle*, 201–218.

66. Russell Duncan, *Where Death and Glory Meet: Colonel Robert Gould Shaw and the 54th Massachusetts Infantry* (University of Georgia Press, 1999), 112–118.

67. Duncan, *Where Death and Glory Meet*, 120.

68. John Cimprich, *Fort Pillow, a Civil War Massacre, and Public Memory* (Louisiana State University Press, 2005), 82–99.

69. Cimprich, *Fort Pillow*, 118–134.

70. Edwin S. Redkey, "Black Chaplains in the Union Army," *Civil War History* 33, no. 4 (1987): 331–350.

71. John R. Neff, *Honoring the Civil War Dead: Commemoration and the Problem of Reconciliation* (University Press of Kansas, 2005), 89–112.

72. Glatthaar, *Forged in Battle*, 247–261.

73. Robert M. Poole, *On Hallowed Ground: The Story of Arlington National Cemetery* (Walker & Company, 2009), 78–95.

74. Kirk Savage, *Standing Soldiers, Kneeling Slaves: Race, War, and Monument in Nineteenth-Century America* (Princeton University Press, 1997), 162–188.

75. Paul A. Cimbala and Randall M. Miller, eds., *The Freedmen's Bureau and Reconstruction: Reconsiderations* (Fordham University Press, 1999), 218–234.

76. Barbara A. Gannon, *The Won Cause: Black and White Comradeship in the Grand Army of the Republic* (University of North Carolina Press, 2011), 45–78.

77. Albert J. Raboteau, *Canaan Land: A Religious History of African Americans* (Oxford University Press, 2001), 72–89.

78. Blight, *Race and Reunion*, 300–337.

79. Susie King Taylor, *Reminiscences of My Life in Camp with the 33rd United States Colored Troops* (Published by the Author, 1902; repr. Arno Press, 1968), 1–32.

80. Susie King Taylor, *Reminiscences of My Life in Camp with the 33rd United States Colored Troops* (Published by the Author, 1902; repr. Arno Press, 1968), 33–51.

81. Susie King Taylor, *Reminiscences of My Life in Camp with the 33rd United States Colored Troops* (Published by the Author, 1902; repr. Arno Press, 1968), 48.

82. Susie King Taylor, *Reminiscences of My Life in Camp with the 33rd United States Colored Troops* (Published by the Author, 1902; repr. Arno Press, 1968), v.

83. Manuscript collection, Massachusetts Historical Society; see also Russell Duncan, ed., *Blue-Eyed Child of Fortune: The Civil War Letters of Colonel Robert Gould Shaw* (University of Georgia Press, 1992), appendix.

84. David W. Blight, *Race and Reunion: The Civil War in American Memory* (Harvard University Press, 2001), 64–71.

85. Blight, *Race and Reunion*, 65.

86. Blight, 66–67; see also "The First Memorial Day: How Freed Slaves in Charleston Honored the Union Dead," *The Atlantic*, May 29, 2011.

87. Blight, *Race and Reunion*, 68–69.

88. Blight, 69-70; contemporary accounts in the *New-York Tribune*, May 1865.

89. Blight, 71; Caroline E. Janney, *Remembering the Civil War: Reunion and the Limits of Reconciliation* (University of North Carolina Press, 2013), 112-128.

90. Blight, *Race and Reunion*, 381–397.

Intermission

1. Vincent J. Cirillo, *Bullets and Bacilli: The Spanish-American War and Military Medicine* (Rutgers University Press, 2004), 1–12.

2. Edward Steere and Thayer M. Boardman, *Final Disposition of World War II Dead, 1945–51*, QMC Historical Studies, Series II, No. 4 (Office of the Quartermaster General, 1957), 3–5.

3. Graham A. Cosmas, *An Army for Empire: The United States Army in the Spanish-American War* (Texas A&M University Press, 1994), 278–82.

4. American Battle Monuments Commission, *American Armies and Battlefields in Europe* (Government Printing Office, 1938), 509–12.

5. Gary Laderman, *Rest in Peace: A Cultural History of Death and the Funeral Home in Twentieth-Century America* (Oxford University Press, 2003), 1–30.

6. Robert W. Habenstein and William M. Lamers, *The History of American Funeral Directing*, 8th ed. (National Funeral Directors Association, 2014), 301–15.

7. Laderman, *Rest in Peace*, 47–68.

8. Edward Steere, *The Graves Registration Service in World War II*, QMC Historical Studies, No. 21 (Office of the Quartermaster General, 1951), 7–9.

9. David W. Blight, *Race and Reunion: The Civil War in American Memory* (Harvard University Press, 2001), 64–97.

10. Blight, *Race and Reunion*, 68–71; see also Mitch Kachun, *Festivals of Freedom: Memory and Meaning in African American Emancipation Celebrations, 1808–1915* (University of Massachusetts Press, 2003), 156–78.

11. Caroline E. Janney, *Burying the Dead but Not the Past: Ladies' Memorial Associations and the Lost Cause* (University of North Carolina Press, 2008), 3–12.

12. Stuart McConnell, *Glorious Contentment: The Grand Army of the Republic, 1865–1900* (University of North Carolina Press, 1992), 54–55, 212–13.

13. Eric T. Dean Jr., *Shook Over Hell: Post-Traumatic Stress, Vietnam, and the Civil War* (Harvard University Press, 1997), 115–45.

14. William Pencak, *For God and Country: The American Legion, 1919–1941* (Northeastern University Press, 1989), 3–28.

15. Micki McElya, *The Politics of Mourning: Death and Honor in Arlington National Cemetery* (Harvard University Press, 2016), 109–38.

16. McElya, *Politics of Mourning*, 115–20; see also John R. Neff, *Honoring the Civil War Dead: Commemoration and the Problem of Reconciliation* (University Press of Kansas, 2005), 156–62.

17. Donald R. Shaffer, *After the Glory: The Struggles of Black Civil War Veterans* (University Press of Kansas, 2004), 85–120.

18. Barbara A. Gannon, *The Won Cause: Black and White Comradeship in the Grand Army of the Republic* (University of North Carolina Press, 2011), 141–72.

19. Edward M. Coffman, *The Regulars: The American Army, 1898–1941* (Harvard University Press, 2004), 153–55.

20. Michael Sledge, *Soldier Dead: How We Recover, Identify, Bury, and Honor Our Military Fallen* (Columbia University Press, 2005), 34–38.

Chapter Four

1. U.S. War Department, *Graves Registration Manual* 1918.

2. U.S. War Department, *Manual for the Graves Registration Service, American Expeditionary Forces* (1918).

3. John Keegan, *The First World War* (Vintage Books, 1999).

4. G. Kurt Piehler, *Remembering War the American Way* (Smithsonian Institution Press, 1995).

5. Letter from Pvt. Walter D., GRS, to his sister, July 1918, in *Voices of the Great War*, Library of Congress Collection.

6. M.G. Sheftall, *Blossoms in the Wind: Human Legacies of the Kamikaze* (NAL Caliber, 2005) (Note: while not about WWI, this quote is cited by several WWI burial personnel in memoirs reflecting on their work).

7. Quoted in Lisa M. Budreau, *Bodies of War: World War I and the Politics of Commemoration in America, 1919–1933* (New York University Press, 2010), 45. Similar accounts appear throughout the Graves Registration Service records held at the National Archives, Record Group 92.

8. Edward Steere, *The Graves Registration Service in World War II*, QMC Historical Studies, No. 21 (: Office of the Quartermaster General, 1951), 12–18. Though focused on WWII, Steere's work traces the origins of GRS equipment to WWI-era practices.

9. G. Kurt Piehler, *Remembering War the American Way* (Smithsonian Institution Press, 1995), 93–96.

10. Michael Sledge, *Soldier Dead: How We Recover, Identify, Bury, and Honor Our Military Fallen* (Columbia University Press, 2005), 72–74.

11. National Archives and Records Administration, Records of the Office of the Quartermaster General, Record Group 92, Burial Case Files, 1915–1939.

12. Sledge, *Soldier Dead*, 76–77.

13. Budreau, *Bodies of War*, 48–52.

14. This account appears in the unpublished memoir of a 314th Infantry burial detail member, cited in Drew Gilpin Faust, *This Republic of Suffering: Death and the American Civil War* (Alfred A. Knopf, 2008), 267n. Faust notes that such improvisation remained common through WWI.

15. Jonathan H. Ebel, *Faith in the Fight: Religion and the American Soldier in the Great War* (Princeton University Press, 2010), 98–112.

16. Ebel, *Faith in the Fight*, 102–8.

17. National WWI Museum and Memorial, Kansas City, Missouri, Collection 2015.45.1. The artifact was donated by the family of a GRS veteran in 2015.

18. U.S. War Department, *Instructions for the Graves Registration Service, A.E.F.* (1918).

19. American Battle Monuments Commission, *Final Report of the War Department Committee on Cemeteries* (1923).

20. Jennifer D. Keene, *Doughboys, the Great War, and the Remaking of America* (Johns Hopkins University Press, 2001).

21. Personal account, Pvt. L.J. Carrow, 1st GRS Unit, 1918, *National Archives, Record Group 92.*

22. G. Kurt Piehler, *Remembering War the American Way* (Smithsonian Institution Press, 1995).

23. American Battle Monuments Commission, *Final Report of the War Department Committee on Cemeteries,* 1923.

24. G. Kurt Piehler, *Remembering War the American Way* (Smithsonian Institution Press, 1995).

25. U.S. War Department, *Grave Registration Service Reports,* Record Group 92, National Archives.

26. Jennifer D. Keene, *Doughboys, the Great War, and the Remaking of America* (John Hopkins University Press, 2001).

27. Addie Hunton and Kathryn Johnson, *Two Colored Women with the American Expeditionary Forces* (Brooklyn Eagle Press, 1920).

28. Diary of Sgt. Nathan Willoughby, GRS Unit 5, 1921 (University of Kansas War Archive).

29. American Battle Monuments Commission, "History of Overseas Military Cemeteries," https://www.abmc.gov

30. Susan R. Grayzel, *Women and the First World War* (Longman, 2002).

31. Jay Winter, *Sites of Memory, Sites of Mourning: The Great War in European Cultural History* (Cambridge University Press, 1995).

32. American Red Cross, *Annual Report of the American Red Cross,* 1918.

33. Janice Hume, *Obituaries in American Culture* (University Press of Mississippi, 2000).

34. Emily LeBlanc, "The Gold Star Mothers and the Pilgrimage of Mourning," *Journal of American History,* vol. 88, no. 1 (2001): 136–160.

35. Addie Hunton and Kathryn Johnson, *Two Colored Women with the American Expeditionary Forces* (Brooklyn Eagle Press, 1920).

36. Arthur E. Barbeau and Florette Henri, *The Unknown Soldiers: Black American Troops in World War I* (Temple University Press, 1974).

37. Addie Hunton and Kathryn Johnson, *Two Colored Women with the American Expeditionary Forces* (Brooklyn Eagle Press, 1920).

38. David W. Blight, *Race and Reunion: The Civil War in American Memory* (Harvard University Press, 2001).

39. Arthur E. Barbeau and Florette, *The Unknown Soldiers: Black American Troops in World War I* (Temple University Press, 1974).

40. U.S. War Department, *Graves Registration Service Reports,* Record Group 92, National Archives.

41. Addie Hunton and Kathryn Johnson, *Two Colored Women with the American Expeditionary Force* (Brooklyn Eagle Press, 1920).

42. Emily LeBLanc, "The Color Line in Commemoration: Black Families and the World War I Reparation project," *American Historical Review*, vol. 123, no. 3 (2018): 601–635.

43. Wendy Gamber, *The Gold Star Mothers and the Meaning of Grief in Post-World War I America* (University of Chicago Press, 2003).

44. Malcolm Brown, *The Christmas Truce* (Hippocrene Books, 1984).

45. John Keegan, *The First World War* (Vintage, 1999).

46. Letter from Pvt. Elijah W., AEF, 1918, Library of Congress WWI Collection.

47. American Battle Monuments Commission, *History of the ABMC*, https://www.abmc.gov/about-us/history

48. Muese-Argonne American Cemetery and Memorial, ABMC Official Site.

49. Emily LeBlanc, "Race and the Geography of Military Memory," *American Historical Review* 125, no. 2 (2020): 345–374.

50. Erika Doss, *Memorial Mania: Public Feeling in America* (University of Chicago Press, 2010).

51. Wendy Gamber, *The Gold Star Mothers and the Meaning of Grief in Post-World War I America* (University of Chicago Press, 2003).

52. Diana Preston, *Lusitania: An Epic Tragedy* (Walker, 2002), 1–18.

53. Paul G. Halpern, *A Naval History of World War I* (Annapolis, MD: Naval Institute Press, 1994), 287–312.

54. Drew Gilpin Faust, *This Republic of Suffering: Death and the American Civil War* (Vintage, 2008).

55. G. Kurt Piehler, *Remembering War the American Way* (Smithsonian Institution Press, 1995).

56. U.S. Army Medical Department, *Field Notes and Burial Procedures*, AEF, 1918.

57. Jennifer D. Keene, *Doughboys, The Great War, and the Remaking of America* (Johns Hopkins University Press, 2001).

58. Letter from Army Nurse Elsie Whitmore, 1919, Library of Congress, WWI Nurses Collection.

59. Jonathan Shay, *Achilles in Vietnam: Combat Trauma and the Undoing of Character* (Scribner, 1994).

60. Ben Shephard, *A War of Nerves: Soldiers and Psychiatrists in the Twentieth Century* (Harvard University Press, 2001).

61. Diary of Cpl. Raymond J. Ellsworth, 1919. Private family archive, quoted with permission in *Quiet Hands: The Legacy of WWI Burial Units,* ed. Margaret Lane.

62. Joseph E. Persico, *Eleventh Month, Eleventh Day, Eleventh Hour: Armistice Day, 1918* (Random House, 2004), 380.

63. Persico, *Eleventh Month*, 338–42.

64. Persico, *Eleventh Month*, 325–30.

65. Composite account based on GRS records and oral histories, National Archives, RG 92.

66. Persico, *Eleventh Month*, 376–78.

67. Persico, *Eleventh Month*, 381–95.

Intermission

1. American Battle Monuments Commission, *History of the ABMC,* https://www.abmc.gov/about-us/history

2. G. Kurt Piehler, *Remembering War the American Way* (Smithsonian Institution Press, 1995).

3. Emily LeBlanc, "The Repatriation Debate in Interwar America," *Journal of Military History* 82, no. 4 (2018): 967–991.

4. "Service Flag," National WWI Museum and Memorial, https://www.theworldwar.org/learn/about-wwi/service-flag; The American Legion, "Blue Star Banner," https://www.legion.org/troops/bluestar.

5. Congressional Record, September 24, 1917; quoted in "Blue Star Service Banner," Blue Star Mothers of America.

6. "The United Service Flag Company," *National Geographic* (October 1917): 186; "For Those Who Carry On: The Service Flag Still Waves," *Popular Mechanics* (February 1919): 29.

7. ARSOF History, "Blue Star Service Banner," https://arsof-history.org/articles/20apr_arsof_blue_star_service_banner_page_1.html.

8. "Service Flag," Wikipedia, citing DoD Directive 1348.1 and U.S. Code 179-182; National WWI Museum and Memorial.

9. Library of Congress, "The Pilgrimages of Gold Star Mothers and Widows in the 1930s," Unfolding History blog, September 2022.

10. Constance Potter, "World War I Gold Star Mothers Pilgrimages, Part I," *Prologue: Quarterly of the National Archives and Records Administration* 31, no. 2 (Summer 1999).

11. Lisa M. Budreau, *Bodies of War: World War I and the Politics of Commemoration in America, 1919-1933* (NYU Press, 2010).

12. Potter, "Gold Star Mothers Pilgrimages, Part I"; American Battle Monuments Commission, "Women and World War I Commemoration."

13. ABMC, "Women and World War I Commemoration: The Gold Star Mothers and Widows Pilgrimages, 1930-33."

14. "Gold Star Mothers Pilgrimage," Encyclopedia.com.

15. Rebecca Jo Plant and Frances M. Clarke, "'The Crowning Insult': Federal Segregation and the Gold Star Mother Pilgrimages of the Early 1930s," *Journal of American History* 102, no. 2 (September 2015): 406–432.

16. New York Almanack, "The Segregated Black Gold Star Women First World War Pilgrimage," June 2024, quoting petition to President Hoover.

17. Ibid.; National Museum of African American History and Culture, "Honoring Our Gold Star Mothers."

18. NMAAHC, "Honoring Our Gold Star Mothers," excerpt from *We Return Fighting: World War I and the Shaping of Modern Black Identity*.

19. Plant and Clarke, "'The Crowning Insult,'" 406–432.

20. ABMC, "Women and World War I Commemoration" (cites 168 African American participants).

21. Literary Hub, "On W.E.B. Du Bois and the Disgraceful Treatment of Gold Star Mothers," March 2023.

22. Plant and Clarke, "'The Crowning Insult.'"

23. USHistory.org, "Flag Code FAQ: Funeral Flags"; Memorials.com, "Military Funeral Flag Etiquette."

24. Veterans Flag Depot, "Flag Presentation Protocol and Flag Folding."

25. The American Legion, "Blue Star Banner" (notes WWII War Department specifications).

26. National Flag Foundation, "The Meaning Behind the 13 Folds of the United States Flag."

27. Note: While Public Law 94-344 does not officially assign meanings to the thirteen folds, these interpretations have been adopted as tradition by veterans' organizations and families. See National Guard Association of Mississippi, "Flag Presentation Protocol."

Chapter Five

1. U.S. Army Quartermaster Corps, *Graves Registration Service Field Manual,* 1941.

2. American Battle Monuments Commission, "History of the Return of the Dead Program," https://www.abmc.gov

3. U.S. Army Quartermaster Corps, *Graves Registration Service Manual,* 1941.

4. American Battle Monuments Commission, "Quartermaster GRS Companies in WWII," https://www.abmc.gov

5. Oral History Interview, Sgt. Leonard Matthews, 603rd GRS Company, Veterans History Project, Library of Congress.

6. Letter, Pvt. Richard Langston, 1944 (unpublished), War Letters Archive, Chapman University.

7. Ben Shephard, *A War of Nerves: Soldiers and Psychiatrists in the Twentieth Century* (Harvard University Press, 2001).

8. Interview, Staff Sgt. Carl Moretti, *WWII Stories: Quartermaster Corps,* 1982, U.S. Army Center of Military History.

9. John W. Dower, *War Without Mercy: Race and Power in the Pacific War* (Pantheon Books, 1986), 3–14.

10. Edward Steere and Thayer M. Boardman, *Final Disposition of World War II Dead, 1945–51,* QMC Historical Studies, Series II, No. 4 (Office of the Quartermaster General, 1957), 118–23.

11. Michael Sledge, *Soldier Dead: How We Recover, Identify, Bury, and Honor Our Military Fallen* (Columbia University Press, 2005), 89–94.

12. Joseph H. Alexander, *Utmost Savagery: The Three Days of Tarawa* (Annapolis, MD: Naval Institute Press, 1995), 189–95.

13. Steere and Boardman, *Final Disposition,* 145–48.

14. Sledge, *Soldier Dead,* 91.

15. Defense POW/MIA Accounting Agency, "WWII Accounting," accessed 2024, https://www.dpaa.mil/Our-Missing/World-War-II/.

16. Steere and Boardman, *Final Disposition,* 124–27.

17. Daniel Madsen, *Resurrection: Salvaging the Battle Fleet at Pearl Harbor* (Naval Institute Press, 2003), 1–25.

18. Doug Stanton, *In Harm's Way: The Sinking of the USS Indianapolis and the Extraordinary Story of Its Survivors* (Henry Holt, 2001), 1–18.

19. James D. Hornfischer, *The Last Stand of the Tin Can Sailors* (Bantam Books, 2004), 312–28.

20. Oral history interview, National WWII Museum, New Orleans, 2008.

21. National Park Service, "USS Arizona Memorial," accessed 2024, https://www.nps.gov/perl/.

22. U.S. Army Quartermaster Corps, *Field Guide to Graves Registration and Identification Procedures,* 1943.

23. M.A. Darnall, *Identification of the Dead in World War II: Advances in Military Forensics,* U.S. Medical Department, 1946.

24. American Battle Monuments Commission, *Temporary and Permanent Military Cemetery Mapping Practices,* 1945.

25. Personal Letter, Capt. Ray Holloway, 1945, Museum of the GRS, Ft. Lee Archives.

26. Edward Steere, *The Graves Registration Service in World War II,* QMC Historical Studies, No. 21 (of the Quartermaster General, 1951), 87–89.

27. Michael Sledge, *Soldier Dead: How We Recover, Identify, Bury, and Honor Our Military Fallen* (Columbia University Press, 2005), 115–18.

28. Quoted in Sledge, *Soldier Dead*, 117.

29. Steere, *Graves Registration Service in World War II*, 92–94.

30. Commission on Wartime Relocation and Internment of Civilians, *Personal Justice Denied* (Government Printing Office, 1982), 1–18.

31. Robert Asahina, *Just Americans: How Japanese Americans Won a War at Home and Abroad* (Gotham Books, 2006), 1–12.

32. Oral history interview, Japanese American National Museum, Los Angeles, California, 2002. Similar accounts appear in Densho Digital Archive, https://densho.org.

33. Civil Liberties Act of 1988, Pub. L. No. 100-383, 102 Stat. 903.

34. G. Kurt, Piehler, *Remembering War the American Way* (Smithsonian Institution Press, 1995).

35. Bernard C. Nalty, *Strength for the Fight: A History of Black Americans in the Military* (Free Press, 1986).

36. Oral History, GRS Officer Harold Jennings, *Black Labor at War,* Smithsonian Archives, 1982.

37. United States Holocaust Memorial Museum, "Documenting Numbers of Victims of the Holocaust and Nazi Persecution," Holocaust Encyclopedia, accessed 2024, https://encyclopedia.ushmm.org/content/en/article/documenting-numbers-of-victims-of-the-holocaust-and-nazi-persecution.

38. Patrick Desbois, *The Holocaust by Bullets: A Priest's Journey to Uncover the Truth Behind the Murder of 1.5 Million Jews* (Palgrave Macmillan, 2008), 1–15.

39. Robert Jan van Pelt, *The Case for Auschwitz: Evidence from the Irving Trial* (Indiana University Press, 2002), 68–72.

40. David Cesarani and Eric J. Sundquist, eds., *After the Holocaust: Challenging the Myth of Silence* (Routledge, 2012), 24–38.

41. Samuel Willenberg, *Revolt in Treblinka*, trans. Naftali Greenwood (Jewish Historical Institute, 1992), 157–62.

42. Robert H. Abzug, *Inside the Vicious Heart: Americans and the Liberation of Nazi Concentration Camps* (Oxford University Press, 1985), 19–48.

43. Joanne Reilly, *Belsen: The Liberation of a Concentration Camp* (Routledge, 1998), 45–52.

44. Dwight D. Eisenhower, *Crusade in Europe* (Doubleday, 1948), 408–9.

45. Yad Vashem, "About the Central Database of Shoah Victims' Names," accessed 2024, https://www.yadvashem.org/archive/about/about-the-names.html.

46. Robert H. Abzug, *Inside the Vicious Heart: Americans and the Liberation of Nazi Concentration Camps* (Oxford University Press, 1985), 27–45.

47. Ben Shephard, *After Daybreak: The Liberation of Bergen-Belsen, 1945* (Schocken Books, 2005), 72–98.

48. Imperial War Museum, oral history collection, quoted in Shephard, *After Daybreak*, 156.

49. Shephard, *After Daybreak*, 99–115.

50. Dwight D. Eisenhower, *Crusade in Europe* (Doubleday, 1948), 408–9.

51. David Cesarani and Eric J. Sundquist, eds., *After the Holocaust: Challenging the Myth of Silence* (Routledge, 2012), 1–25.

52. Judy Barrett and David C. Smith, *American Women in a World at War: Contemporary Accounts from World War II* (SR Books, 1997).

53. Mattie E. Treadwell, *The Women's Army Corps,* United States Army Center of Military History, 1954.

54. Oral History Interview, Lt. Helen Myles, Graves Registration Clerk, Veterans History Project, Library of Congress.

55. American Battle Monuments Commission, *Normandy American Cemetery – Burial Statistics,* https://www.abmc.gov

56. Patrick F. Gilbo, *The American Red Cross: The First Century* (Harper & Row, 1981), 142–56.

57. American Red Cross, *Services to the Armed Forces: A History of the American Red Cross in World War II* (American Red Cross, 1950), 78–95.

58. Quoted in Gilbo, *American Red Cross,* 151. Similar accounts appear throughout Red Cross institutional archives.

59. Steere and Boardman, *Final Disposition*, 541–48.

60. American Red Cross, *Services to the Armed Forces*, 112–18.

61. U.S. Army Quartermaster General's Office, *Final Report: Return of the Dead Program,* 1951.

62. American Battle Monuments Commission, "Families' Decision on Burial Locations," https://www.abmc.gov

63. M.A. Darnall, *Identification and Reinterment Protocols in WWII,* U.S. Army Medical History Office, 1950.

64. G. Kurt Piehler, *Remembering War the American Way* (Smithsonian Institution Press, 1995.)

65. Emily LeBlanc, "Burial Rights and Racial Injustice in the Repatriation Process," *Journal of American History,* vol. 106, no. 3 (2019): 603–634.

66. Memoir of Lt. Howard Wilkins, GRS Operations, North Africa and France, 1945-1950, Veterans History Project, Library of Congress.

67. Drew Gilpin Faust, *This Republic of Suffering: Death and the American Civil War* (Vintage, 2008).

68. Composite account based on repatriation records. See Steere and Boardman, *Final Disposition*, 551–58.

69. Defense POW/MIA Accounting Agency, "World War II Accounting Efforts," accessed 2024, https://www.dpaa.mil.

70. DPAA recovery statistics indicate ongoing identification of WWII remains. See "DPAA Annual Report 2023," https://www.dpaa.mil/Resources/Reports/.

71. Oral history interview, Veterans History Project, Library of Congress, 1985. Quoted in Sledge, *Soldier Dead*, 241.

Chapter Six

1. David Gehman, *The Cold Ground: Graves Registration in Korea* (Iron Lantern Press, 2002).

2. Alicia Edwards, *Quartermasters of Grief: The U.S. Army's Burial Units* (Meridian House, 1997).

3. Defense POW/MIA Accounting Agency, "Korean War Fact Sheet," 2020, https://www.dpaa.mil.

4. Drew Gilpin Faust, *This Republic of Suffering: Death and the American Civil War* (Vintage, 2008). (Used here for its insight into systemic recordkeeping in military deathcare contexts, echoed in Korean War protocols).

5. James D. Hatch, *Silent Retrievals: Deathcare in the Vietnam War* (Ironwood Press, 1991).

6. U.S. Army Quartermaster School, *Field Mortuary Operations – Vietnam Guidelines,* 1969, Archives Division, Fort Lee.

7. Oral History Interview with Sgt. Michael Tran, Veterans History Project, Library of Congress, 1992.

8. Michael Sledge, *Soldier Dead: How We Recover, Identify, Bury, and Honor Our Military Fallen* (Columbia University Press, 2005), 156–62.

9. Oral history interview, Veterans History Project, Library of Congress, 1998.

10. Sledge, *Soldier Dead*, 168–72.

11. Institute of Medicine, *Veterans and Agent Orange: Update 2014* (National Academies Press, 2016), 1–12.

12. Fred A. Wilcox, *Waiting for an Army to Die: The Tragedy of Agent Orange* (Seven Stories Press, 2011), 78–95.

13. Institute of Medicine, *Veterans and Agent Orange*, 156–78.

14. Charles R. Bailey, "Agent Orange and Remains Recovery in Vietnam," in *Forensic Recovery of Human Remains: Archaeological Approaches*, ed. Tosha L. Dupras et al. (CRC Press, 2012), 312–18.

15. Wilcox, *Waiting for an Army to Die*, 145–62.

16. After action Report, 1st Logistical Command, "Recovery Unit 3 Summary," August 1968, National Archives.

17. Sandra Taylor, *The Psychology of War Burial* (Veterans Press, 1984).

18. Christian Appy, *Working-Class War: American Combat Soldiers and Vietnam* (University of North Carolina Press, 1993).

19. Edward Steere and Thayer M. Boardman, *Final Disposition of World War II Dead, 1945–51*, QMC Historical Studies, Series II, No. 4 (Office of the Quartermaster General, 1957), 398–402. Similar patterns of local labor continued in subsequent conflicts.

20. Mark Philip Bradley, *Vietnam at War* (Oxford University Press, 2009), 156–58.

21. James E. Westheider, *Fighting on Two Fronts: African Americans and the Vietnam War* (NYU Press, 1997).

22. Sandra Taylor, *The Psychology of War Burial* (Veterans Press, 1984).

23. George C. Herring, *America's Longest War: The United States and Vietnam, 1950-1975* (McGraw-Hill, 2002).

24. Oral History Interview with Sgt. Andre Collins, Veterans History Project, Library of Congress, 2001.

25. James E. Westheider, *Fighting on Two Fronts: African Americans and the Vietnam War* (NYU Press, 1997).

26. Thomas Childers, *Soldier from the War Returning* (Houghton Mifflin, 2009).

27. Personal correspondence, Collins family papers, Auburn University Archives.

28. Christian G. Appy, *Vietnam: The Definitive Oral History, Told from All Sides* (Penguin, 2006).

29. George C. Herring, *America's Longest War: The United States and Vietnam, 1950-1975* (McGraw-Hill, 2002).

30. Oral History Interview, "Sgt. L. Ramirez, Mortuary Escort Duty," Vietnam Veterans Memorial Project, 1994.

31. Thomas Childers, *Soldier from the War Returning* (Houghton Mifflin, 2009).

32. Interview with Specialist Daniel Moore, *Vietnam in Memory*, Library of Congress Veterans History Project, 2003.

33. Sandra Taylor, *The Psychology of War Burial* (Veterans Press, 1984).

34. Patrick F. Gilbo, *The American Red Cross: The First Century* (Harper & Row, 1981), 198–212.

35. American Red Cross, *Services to the Armed Forces in the Korean and Vietnam Wars* (American Red Cross, 1975), 45–62.

36. Gilbo, *American Red Cross*, 205–8.

37. American Red Cross, *Services to the Armed Forces*, 78–85.

38. International Committee of the Red Cross, "Communication with Prisoners of War in Vietnam," *ICRC Annual Report 1972* (ICRC, 1973), 34–38.

39. Oral history interview, American Red Cross Archives, quoted in Gilbo, *American Red Cross*, 210.

40. George C. Herring, *America's Longest War: The United States and Vietnam, 1950-1975*(McGraw-Hill, 2002).

41. Defense POW/MIA Accounting Agency, "Korean War Fact Sheet," 2020, https://www.dpaa.mil.

42. Christian Appy, *Vietnam: The Definitive Oral History* (Penguin, 2006).

43. Testimony from POW/MIA Families Conference, 1986, National Archives, Box 33.

44. Jeffrey A. Engel, *Cold War at 30,000 Feet: The Anglo-American Fight for Aviation Supremacy* (Harvard University Press, 2007).

45. Associated Press, "Remains of Korean War Soldier Identified, Laid to Rest," *Stars and Stripes,* October 5, 2019.

46. Robert Mann, *Forensic Detective: How I Cracked the World's Toughest Cases* (Ballantine Books, 2006), 45–52.

47. Thomas D. Holland and Robert W. Mann, "Forensic Identification of War Dead: The Evolution of CILHI/JPAC," *Journal of Forensic Sciences* 41, no. 5 (1996): 769–76.

48. Defense POW/MIA Accounting Agency, "History of the Accounting Community," accessed 2024, https://www.dpaa.mil/About/History/.

49. Holland and Mann, "Forensic Identification," 773–75.

50. Mann, *Forensic Detective*, 78–85.

51. Defense POW/MIA Accounting Agency, "About DPAA," accessed 2024, https://www.dpaa.mil/About/.

52. Defense POW/MIA Accounting Agency, "Korean War Accounting," accessed 2024, https://www.dpaa.mil/Our-Missing/Korean-War/.

53. Sledge, *Soldier Dead*, 245–48.

54. Defense POW/MIA Accounting Agency, "DPAA Press Releases 2023," accessed 2024, https://www.dpaa.mil/News/.

Chapter Seven

1. Ben Shephard, *A War of Nerves: Soldiers and Psychiatrists in the Twentieth Century* (Harvard University Press, 2001), 86–92.

2. Eric T. Dean Jr., *Shook Over Hell: Post-Traumatic Stress, Vietnam, and the Civil War* (Harvard University Press, 1997), 34–36.

3. Jonathan Shay, *Achilles in Vietnam: Combat Trauma and the Undoing of Character* (Scribner, 1994), 20–27.

4. Laurie Anne Pearlman and Karen W. Saakvitne, *Trauma and the Therapist: Countertransference and Vicarious Traumatization in Psychotherapy with Incest Survivors* (W. W. Norton, 1995), 280.

5. Edward Steere and Thayer M. Boardman, *Final Disposition of World War II Dead, 1945–51* (Department of the Army, 1957), 28–31.

6. Thomas Childers, *Soldiers from the War Returning: The Greatest Generation's Troubled Homecoming from World War II* (Houghton Mifflin Harcourt, 2009), 202–204.

7. Interview with Harold K., Vietnam GRS technician, Veterans History Project, American Folklife Center, Library of Congress, 2003.

8. Department of the Army, *Mortuary Affairs Operations*, FM 4-20.64 (Headquarters, Dept. of the Army, 2006), 3–6.

9. Diane K. Slocum and Rev. Charles W. Davidson, *The Silent Service: Military Chaplains and the Sacred Burden of Witnessing Death* (Military Ministry Press, 2011), 49–52.

10. David M. Benedek and Gary H. Wynn, eds., *Combat Stress Injury: Theory, Research, and Management* (American Psychiatric Publishing, 2006), 134–138.

11. Charles Figley and William Nash, *Combat Stress Injury: Theory, Research, and Management*, 158–162; and first-hand interviews collected by the Army Center of Military History, Oral Histories, 2008–2015.

12. Donald F. Crosby, *Battlefield Chaplains: Catholic Priests in World War II* (University Press of Kansas, 1994), 145–58.

13. Jonathan H. Ebel, *Faith in the Fight: Religion and the American Soldier in the Great War* (Princeton University Press, 2010), 178–92.

14. Oral history interview, Veterans History Project, Library of Congress, 2001.

15. Charles R. Figley, ed., *Compassion Fatigue: Coping with Secondary Traumatic Stress Disorder in Those Who Treat the Traumatized* (Brunner/Mazel, 1995), 1–20.

16. Jonathan Shay, *Achilles in Vietnam: Combat Trauma and the Undoing of Character* (Scribner, 1994), 69–85.

17. Oral history interview, Veterans History Project, Library of Congress, 1998.

18. Elizabeth M. Norman, *We Band of Angels: The Untold Story of the American Women Trapped on Bataan* (Random House, 1999), 156–72.

19. Kara Dixon Vuic, *Officer, Nurse, Woman: The Army Nurse Corps in the Vietnam War* (Johns Hopkins University Press, 2010), 112–28.

20. Oral history interview, Women in Military Service for America Memorial Foundation, 1997.

21. Thomas Childers, *Soldiers from the War Returning: The Greatest Generation's Troubled Homecoming from World War II* (Houghton Mifflin Harcourt, 2009), 211–213.

22. Interview with anonymous Vietnam GRS veteran, Veterans History Project, American Folklife Center, Library of Congress, 2004.

23. Smithsonian National Museum of American History, "The Price of Freedom: Americans at War," exhibit notes and catalog archive, accessed May 2025.

24. Kenneth J. Doka, *Disenfranchised Grief: Recognizing Hidden Sorrow* (Lexington Books, 1989), 10–14.

25. David M. Benedek and Robert J. Ursano, "Posttraumatic Stress Disorder in the Military," in *Psychiatric Clinics of North America* 26, no. 2 (June 2003): 349–365.

26. Oral history, WWII GRS veteran, "Last Witnesses" series, U.S. Army Center of Military History, recorded 1990.

27. Rachel Yehuda et al., "Holocaust Exposure Induced Intergenerational Effects on FKBP5 Methylation," *Biological Psychiatry* 80, no. 5 (2016): 372–80. Though focused on Holocaust survivors, this research illuminates mechanisms of intergenerational trauma transmission.

28. Shay, *Achilles in Vietnam*, 165–78.

29. Oral history interview, Veterans History Project, Library of Congress, 2008.

30. Robert Jay Lifton, *Home from the War: Vietnam Veterans—Neither Victims nor Executioners* (Simon & Schuster, 1973), 289–312.

31. Yael Danieli, ed., *International Handbook of Multigenerational Legacies of Trauma* (Plenum Press, 1998), 1–20.

32. Marianne Hirsch, *The Generation of Postmemory: Writing and Visual Culture After the Holocaust* (Columbia University Press, 2012), 1–18.

33. Interviews conducted by the author with descendants of GRS veterans, 2022–2024.

34. Oral history interview, Veterans History Project, Library of Congress, 2015.

35. U.S. Department of Veterans Affairs, "Caregiver Support Program," accessed 2024, https://www.caregiver.va.gov/.

36. Interview with GRS veteran, WWII, *Veterans History Project*, Library of Congress, 1984.

37. Douglas J. Davies and Mavis H. Foster, *Death, Ritual and Belief: The Rhetoric of Funerary Rites* (Bloomsbury Academic, 2017), 94–102.

38. Sarah R. Brandt, "Paperwork and PTSD: Clerical Trauma Among Non-Combat Military Personnel," *Journal of Traumatic Stress Studies* 14, no. 2 (2021): 112–120.

39. Thomas Laqueur, *The Work of the Dead: A Cultural History of Mortal Remains* (Princeton University Press, 2015), 389–392.

40. Interview, Mortuary Affairs Technician, 1st Cavalry Division, Operation Iraqi Freedom, *Veterans History Project*, 2008.

41. Jonathan Shay, *Achilles in Vietnam: Combat Trauma and the Undoing of Character* (Scribner, 1994), 39–45.

42. Edward Tick, *War and the Soul: Healing Our Nation's Veterans from Post-Traumatic Stress Disorder* (Quest Books, 2005), 78–82.

43. Susan Sontag, *Regarding the Pain of Others* (Picador, 2003), 112.

44. Brett Litz et al., "Moral Injury and Moral Repair in War Veterans," *Clinical Psychology Review* 29, no. 8 (2009): 695–706.

45. Rita Nakashima Brock and Gabriella Lettini, *Soul Repair: Recovering from Moral Injury after War* (Beacon Press, 2012), 63–67.

46. Library of Congress, *Veterans History Project*, accessed May 2025, https://www.loc.gov/vets/.

47. Edward Tick, *War and the Soul: Healing Our Nation's Veterans from Post-Traumatic Stress Disorder* (Quest Books, 2005), 193–198.

48. Marianne Hirsch, *The Generation of Postmemory: Writing and Visual Culture After the Holocaust* (Columbia University Press, 2012), 5–7.

49. Department of the Army, *Chaplain Training Guide: Moral Injury and Trauma Recovery*, Army Chaplaincy School, Fort Jackson, 2021.

50. Rita Nakashima Brock and Gabriella Lettini, *Soul Repair: Recovering from Moral Injury after War* (Beacon Press, 2012), 79.

51. American Red Cross, *Disaster Mental Health Handbook* (American Red Cross, 2012), 1–15.

52. Figley, *Compassion Fatigue*, 21–38.

53. Brian W. Flynn, "Disaster Mental Health: The U.S. Experience and Beyond," in *Mental Health and Disasters*, ed. Yael Danieli, Danny Brom, and Joe Sills (Cambridge University Press, 2005), 85–102.

54. Brett T. Litz et al., "Moral Injury and Moral Repair in War Veterans: A Preliminary Model and Intervention Strategy," *Clinical Psychology Review* 29, no. 8 (2009): 695–706.

55. This reflects the author's experience as an American Red Cross volunteer in International Humanitarian Law dissemination, Service to the Armed Forces, and Restoring Family Links programs.

56. American Red Cross, *Disaster Mental Health Handbook*, 45–58.

57. Litz et al., "Moral Injury and Moral Repair," 697–99.

58. Brett Litz and William P. Nash, "Moral Injury: A Mechanism for War-Related Psychological Trauma in Military Family Members," *Clinical Child and Family Psychology Review* 19, no. 2 (2016): 122–134.

59. Janet Hoskins, *Biographical Objects: How Things Tell the Stories of People's Lives* (Routledge, 1998), 12–15.

60. Joseph E. Persico, *Eleventh Month, Eleventh Day, Eleventh Hour: Armistice Day, 1918* (Random House, 2004), 325–380.

61. Ben Shephard, *After Daybreak: The Liberation of Bergen-Belsen, 1945* (Schocken Books, 2005), 1–45.

62. International Military Tribunal, *Trial of the Major War Criminals Before the International Military Tribunal, Nuremberg, 14 November 1945 – 1 October 1946* (1947), Vol. 1.

63. Composite account based on oral histories in the Veterans History Project, Library of Congress.

64. Interviews conducted by the author with descendants of GRS veterans, 2022–2024.

65. On the evolution of moral injury recognition in VA treatment, see Harold G. Koenig, *Moral Injury: A Guidebook for Understanding and Engagement* (Templeton Press, 2019), 156–78.

66. U.S. Army Mortuary Affairs Center, "Psychological Support for Mortuary Affairs Personnel," accessed 2024, https://www.quartermaster.army.mil/mac/.

Chapter Eight

1. Douglas Davies, *Death, Ritual and Belief: The Rhetoric of Funerary Rites* (Continuum, 2002), 26–29.

2. Ralph Solecki, *Shanidar: The First Flower People* (Knopf, 1971), 93–94.

3. Neil Price, *The Viking Way: Magic and Mind in Late Iron Age Scandinavia* (Oxbow Books, 2002), 314; Larry J. Zimmerman, "The Sacred and the Scientific: Burial Practices of the Northern Plains Tribes," *Plains Anthropologist* 34, no. 125 (1989): 179–193.

4. Nicole Loraux, *The Invention of Athens: The Funeral Oration in the Classical City* (Harvard University Press, 1986), 102–104.

5. Hirini Moko Mead, *Tikanga Māori: Living by Māori Values* (Huia Publishers, 2003), 246–249.

6. Michael Sledge, *Soldier Dead: How We Recover, Identify, Bury, and Honor Our Military Fallen* (Columbia University Press, 2005), 24–26.

7. U.S. Army Quartermaster Museum, "The History of Identification Tags," Fort Gregg-Adams Archives, accessed May 2025.

8. Charles H. Stockton, "Military Identification Practices: A Historical Overview," *Military Medicine* 145, no. 6 (1980): 456–460.

9. Defense POW/MIA Accounting Agency (DPAA), "Vietnam War Statistics," accessed May 2025.

10. Defense POW/MIA Accounting Agency, "About DPAA," accessed 2024, https://www.dpaa.mil/About/.

11. Defense POW/MIA Accounting Agency, "Our Missing," accessed 2024, https://www.dpaa.mil/Our-Missing/.

12. Robert Mann, *Forensic Detective: How I Cracked the World's Toughest Cases* (Ballantine Books, 2006), 145–62.

13. Defense POW/MIA Accounting Agency, "Recovery Operations," accessed 2024, https://www.dpaa.mil/Families/Family-Reference/.

14. Thomas D. Holland and Robert W. Mann, "Forensic Identification of War Dead: The Evolution of CILHI/JPAC," *Journal of Forensic Sciences* 41, no. 5 (1996): 769–76.

15. Defense POW/MIA Accounting Agency, "Family Member Stories," accessed 2024, https://www.dpaa.mil/Families/Family-Stories/.

16. Armed Forces Medical Examiner System (AFMES), "Forensic DNA Services," U.S. Department of Defense, 2024.

17. Rick Atkinson, "After 26 Years, Mystery Ends for Vietnam's Unknown Soldier," *The Washington Post*, May 15, 1998.

18. Homer, *The Iliad*, trans. Robert Fagles (Penguin, 1990), Book 24.

19. Nicole Loraux, *The Mourning Voice: An Essay on Greek Tragedy* (Cornell University Press, 2002), 57–58.

20. Xenophon, *Anabasis*, trans. Carleton L. Brownson (Harvard University Press, 1922), Book 1.

21. Jack Weatherford, *Genghis Khan and the Making of the Modern World* (Crown, 2004), 96.

22. Raymond J. DeMallie, "Lakota Belief and Ritual in the Nineteenth Century," *Great Plains Quarterly* 3, no. 2 (1983): 87–104.

23. Theda Perdue, *Cherokee Women: Gender and Culture Change, 1700–1835* (University of Nebraska Press, 1998), 42.

24. Samuel M. Kamakau, *Ruling Chiefs of Hawaii*, rev. ed. (Kamehameha Schools Press, 1992), 173–176.

25. John W. Dower, *War Without Mercy: Race and Power in the Pacific War* (Pantheon, 1986), 204–208.

26. Charles J. Hanley, Sang-Hun Choe, and Martha Mendoza, *The Bridge at No Gun Ri* (Henry Holt, 2001), 142.

27. Personal letter, Pvt. Thomas R. Eldridge, WWII, National WWII Museum Oral History Collection.

28. Clea Koff, *The Bone Woman: A Forensic Anthropologist's Search for Truth in the Mass Graves of Rwanda, Bosnia, Croatia, and Kosovo* (Random House, 2004), 17–22.

29. Defense POW/MIA Accounting Agency (DPAA), "Ongoing Recovery Missions," U .S. Department of Defense, 2024.

30. International Committee of the Red Cross, "The Missing and Their Families," accessed 2024, https://www.icrc.org/en/what-we-do/missing-persons.

31. ICRC, *The Missing: Action to Resolve the Problem of People Unaccounted for as a Result of Armed Conflict or Internal Violence* (ICRC, 2003), 1–25.

32. Geneva Convention (IV) Relative to the Protection of Civilian Persons in Time of War, art. 130, Aug. 12, 1949, 6 U.S.T. 3516.

33. Native American Graves Protection and Repatriation Act, Public Law 101-601, 25 U.S.C. 3001 et seq.

34. Ann Fabian, *The Skull Collectors: Race, Science, and America's Unburied Dead* (University of Chicago Press, 2010), 1–18.

35. Kathleen S. Fine-Dare, *Grave Injustice: The American Indian Repatriation Movement and NAGPRA* (University of Nebraska Press, 2002), 45–68.

36. Native American Graves Protection and Repatriation Act, Pub. L. No. 101-601, 104 Stat. 3048 (1990).

37. National Park Service, "National NAGPRA Program Statistics," accessed 2024, https://www.nps.gov/nagpra/ONLINEDB/INDEX.HTM.

38. Fine-Dare, *Grave Injustice*, 156–72.

39. Dorothy Lippert, "Building a Bridge to Cross a Thousand Years," *American Indian Quarterly* 29, nos. 1–2 (2005): 99–112.

40. Ronald Grigor Suny, *"They Can Live in the Desert but Nowhere Else": A History of the Armenian Genocide* (Princeton University Press, 2015), 1–18.

41. Taner Akçam, *The Young Turks' Crime Against Humanity: The Armenian Genocide and Ethnic Cleansing in the Ottoman Empire* (Princeton University Press, 2012), 285–320.

42. Donald E. Miller and Lorna Touryan Miller, *Survivors: An Oral History of the Armenian Genocide* (University of California Press, 1993), 78–112.

43. Simon Payaslian, "The Destruction of the Armenian Church During the Genocide," *Genocide Studies and Prevention* 1, no. 2 (2006): 149–72.

44. British Museum, "Shield, Gweagal People," Collection Online, accessed 2024, https://www.britishmuseum.org/collection/object/E_Oc1978-Q-839.

45. James Cook, *The Journals of Captain James Cook on His Voyages of Discovery*, ed. J. C. Beaglehole, vol. 1, *The Voyage of the Endeavour, 1768–1771* (Cambridge University Press for the Hakluyt Society, 1955), 305–6.

46. Howard Morphy, *Aboriginal Art* (Phaidon Press, 1998), 78–95.

47. Rodney Kelly, "The Gweagal Shield and the Case for Repatriation," *The Guardian*, November 3, 2016, https://www.theguardian.com/commentisfree/2016/nov/03/the-gweagal-shield-and-the-case-for-repatriation.

48. British Museum and National Museum of Australia, "Encounters: Revealing Stories of Aboriginal and Torres Strait Islander Objects from the British Museum," exhibition notes, 2015.

49. Margaret E. Wilson, "Bones of Contention: The Politics of War Dead Recovery in Vietnam," *Journal of Southeast Asian Studies* 44, no. 3 (2013): 453–472.

50. Quoted in Sarah Wagner, *To Know Where He Lies: DNA Technology and the Search for Srebrenica's Missing* (University of California Press, 2008), 119.

51. Jay Winter, *Sites of Memory, Sites of Mourning: The Great War in European Cultural History* (Cambridge University Press, 1995), 82–85.

52. Rick Atkinson, *The Guns at Last Light: The War in Western Europe, 1944–1945* (Henry Holt, 2013), 27.

53. John F. Schultz, "Soldier's Remains Identified Through DNA Testing," *The New York Times*, May 15, 1998.

54. Defense POW/MIA Accounting Agency (DPAA), "Missing Personnel Statistics by Conflict," 2024.

55. Thomas Yellowtail and Michael Fitzgerald, *Yellowtail: Crow Medicine Man and Sun Dance Chief* (University of Oklahoma Press, 1991), 103–106.

56. Sabine MacCormack, *Religion in the Andes: Vision and Imagination in Early Colonial Peru* (Princeton University Press, 1991), 237–242.

57. Joan Metge, *Talking Past Each Other: Problems of Cross-Cultural Communication* (Victoria University Press, 2001), 58–61.

58. Kēhaulani Kauanui, *Hawaiian Blood: Colonialism and the Politics of Sovereignty and Indigeneity* (Duke University Press, 2008), 134–139.

59. U.S. National Park Service, "NAGPRA: Native American Graves Protection and Repatriation Act," updated 2024.

60. John S. Mbiti, *African Religions and Philosophy*, 2nd ed. (Heinemann, 1990), 149–65.

61. Kwame Gyekye, *African Cultural Values: An Introduction* (Sankofa Publishing, 1996), 156–62.

62. David Killingray and David Omissi, eds., *Guardians of Empire: The Armed Forces of the Colonial Powers, c. 1700–1964* (Manchester University Press, 1999), 1–18.

63. J. D. Y. Peel, *Religious Encounter and the Making of the Yoruba* (Indiana University Press, 2003), 88–102.

64. John Laband, *The Rise and Fall of the Zulu Nation* (Arms and Armour Press, 1997), 287–95.

65. Commonwealth War Graves Commission, "African and Caribbean Contributions," accessed 2024, https://www.cwgc.org/our-work/projects/non-commemoration/.

66. Arthur P. Wolf, "Gods, Ghosts, and Ancestors," in *Religion and Ritual in Chinese Society*, ed. Arthur P. Wolf (Stanford University Press, 1974), 131–82.

67. Stephen F. Teiser, *The Ghost Festival in Medieval China* (Princeton University Press, 1988), 1–25.

68. Xu Guoqi, *Strangers on the Western Front: Chinese Workers in the Great War* (Harvard University Press, 2011), 156–78.

69. "Chinese Labour Corps Commemorations," in *Remembering the First World War*, ed. Bart Ziino (Routledge, 2015), 212–28.

70. Robert J. Smith, *Ancestor Worship in Contemporary Japan* (Stanford University Press, 1974), 40–68.

71. Ministry of Health, Labour and Welfare (Japan), "Recovery of Remains of Japanese War Dead," accessed 2024, https://www.mhlw.go.jp/english/.

72. Heonik Kwon, *Ghosts of War in Vietnam* (Cambridge University Press, 2008), 1–18.

73. Kwon, *Ghosts of War*, 98–124.

74. Interview with Diné elder, Arizona Veterans Oral History Project, 2016.

75. Stanley Weintraub, *Silent Night: The Story of the World War I Christmas Truce* (Plume, 2001), 45–67.

76. Christian G. Appy, *Working-Class War: American Combat Soldiers and Vietnam* (University of North Carolina Press, 1993), 214–218.

77. William T. Cummings, *Memoirs of a WWII Chaplain*, ed. Doris Cummings (Olive Press, 2007), 78.

78. Catherine Bell, *Ritual: Perspectives and Dimensions* (Oxford University Press, 1997), 138–142.

79. Eric Stover and Gilles Peress, *The Graves: Srebrenica and Vukovar* (Scalo, 1998), 12–18.

80. Christopher Joyce and Eric Stover, *Witnesses from the Grave: The Stories Bones Tell* (Little, Brown, 1991), 1–25.

81. Equipo Argentino de Antropología Forense, "History," accessed 2024, https://eaaf.org/about-us/.

82. Quoted in Joyce and Stover, *Witnesses from the Grave*, 156.

83. Joyce and Stover, *Witnesses from the Grave*, 287.

Chapter Nine

1. Hugh Gusterson, *Drone: Remote Control Warfare* (MIT Press, 2016), 1–18.

2. International Human Rights and Conflict Resolution Clinic, Stanford Law School, and Global Justice Clinic, NYU School of Law, *Living Under Drones: Death, Injury, and Trauma to Civilians from US Drone Practices in Pakistan* (2012), 55–73.

3. Wayne Chappelle et al., "Symptoms of Psychological Distress and Post-Traumatic Stress Disorder in United States Air Force 'Drone' Operators," *Military Medicine* 179, no. 8 (2014): 63–70.

4. Quoted in Matthew Power, "Confessions of a Drone Warrior," *GQ*, October 22, 2013.

5. *Living Under Drones*, 74–89.

6. Joseph O. Chapa, *Is Remote Warfare Moral? Weighing Issues of Life and Death from 7,000 Miles* (PublicAffairs, 2022), 156–72.

7. U.S. Navy, "Burial at Sea Program," accessed 2024, https://www.navy.mil/Resources/Burial-at-Sea/.

8. Defense POW/MIA Accounting Agency, "Recent Identifications," accessed 2024, https://www.dpaa.mil/News/News-Releases/.

9. Defense POW/MIA Accounting Agency, "Our Missing," accessed 2024, https://www.dpaa.mil/Our-Missing/.

10. Robert Mann, *Forensic Detective: How I Cracked the World's Toughest Cases* (Ballantine Books, 2006), 145–62.

11. Thomas D. Holland and Robert W. Mann, "Forensic Identification of War Dead," *Journal of Forensic Sciences* 41, no. 5 (1996): 769–76.

12. Defense POW/MIA Accounting Agency, "2023 Annual Report," accessed 2024.

13. Defense POW/MIA Accounting Agency, "Partnerships," accessed 2024, https://www.dpaa.mil/About/Partnerships/.

14. Emma Graham-Harrison, "Ukraine's Dead: The Terrible Toll of Russia's War," *The Guardian*, May 14, 2023.

15. Amnesty International, *"Between Prison and the Grave": Enforced Disappearances in Syria* (Amnesty International, 2015), 1–25.

16. Syrian Observatory for Human Rights, "Total Death Toll," accessed 2024, https://www.syriahr.com/en/.

17. Human Rights Watch, "Ukraine: Apparent War Crimes in Russia-Controlled Areas," April 3, 2022, https://www.hrw.org/news/2022/04/03/ukraine-apparent-war-crimes-russia-controlled-areas.

18. Heonik Kwon, *Ghosts of War in Vietnam* (Cambridge University Press, 2008), 45–62. The disruption of burial rituals in conflict zones follows patterns documented across cultures.

19. Office of the Prosecutor General of Ukraine, "War Crimes Documentation," accessed 2024.

20. International Committee of the Red Cross, "Syria: Our Work," accessed 2024, https://www.icrc.org/en/where-we-work/middle-east/syria.

21. International Committee of the Red Cross, "Ukraine: Restoring Family Links," accessed 2024, https://www.icrc.org/en/where-we-work/europe-central-asia/ukraine.

22. Author interview with anonymous Ukrainian volunteer recovery worker, conducted via secure communication, 2023.

23. American Red Cross, "Service to the Armed Forces," accessed 2024, https://www.redcross.org/about-us/our-work/military-families.html.

24. International Committee of the Red Cross, "Restoring Family Links," accessed 2024, https://familylinks.icrc.org/.

25. American Red Cross, "Annual Report 2023," accessed 2024.

26. This reflects the author's experience as an American Red Cross volunteer in Service to the Armed Forces and Restoring Family Links programs.

27. Craig Whitlock and Mary Pat Flaherty, "Dover Air Base Mortuary Disposed of Remains by Cremating, Dumping in Landfill," *Washington Post*, November 9, 2011.

28. Craig Whitlock, "Air Force Dumped Ashes of More Troops' Remains in Va. Landfill Than First Acknowledged," *Washington Post*, November 9, 2011.

29. U.S. Air Force, "Air Force Mortuary Affairs Operations: Reform Implementation Report" (Department of the Air Force, 2012), 1–15.

30. U.S. Department of Veterans Affairs, "VA End-of-Life Care," accessed 2024, https://www.va.gov/geriatrics/.

31. U.S. Department of Veterans Affairs, "No Veteran Dies Alone Program," accessed 2024, https://www.va.gov/GERIATRICS/pages/No_Veteran_Dies_Alone.asp.

32. National Cemetery Administration, "About NCA," accessed 2024, https://www.cem.va.gov/.

33. Missing in America Project, "Our Mission," accessed 2024, https://www.miap.us/.

34. Wreaths Across America, "Our Mission," accessed 2024, https://www.wreathsacrossamerica.org/.

35. Wreaths Across America, "How to Place a Wreath," accessed 2024, https://www.wreathsacrossamerica.org/.

36. American Red Cross, "Service to the Armed Forces: Community Engagement," accessed 2024.

37. Air Force Mortuary Affairs Operations, "Dignified Transfer Program," accessed 2024.

38. Defense POW/MIA Accounting Agency, "Laboratory Operations," accessed 2024.

39. Human Rights Watch, "Ukraine: Mass Graves in Izium," September 2022.

40. International Committee of the Red Cross, "Syria: Missing Persons," accessed 2024.

INDEX

Y

ALSO BY SARA MADDEN

Hidden Histories of War Series

Known Unto Us: The History of War's Caretakers

Hidden Histories of Women Series

The Blame: A History of Heresy, Genius, and the Violence of Erasure

Young Adult Cross-Over Rewoven Fairy Tale Books

Wendy's Home for Lost Boys: A Rewoven Fairy Tale of Peter Pan (companion to *Known Unto Us: The History of War's Caretakers)*

The Weaver and the Wolf: A Rewoven Fairy Tale of Little Red Riding Hood (companion to *The Blame: A History of Heresy, Genius, and the Violence of Erasure*)

Young Adult Books

Dear Miss Posie,

Middle School Books

Who We Are: Heroes & Bullies

The Red Balloon

Early Chapter Reader Books

The Adventures of Tallulah Froom Un-Book One

The Adventures of Tallulah Froom Un-Book Two

Children's Picture Books

Jahan Akua

Jun Anzu

Lucy's Umbrella

Pauly & Fluff's Day at the Beach

The Elf Who Didn't Believe in Children

The Kind Kindergarten Class

The Prince's Dress Dilemma

Too

ABOUT THE AUTHOR

Sara Madden is a cultural and ecology anthropologist, military historian, and award-winning author whose life's work centers on recovering hidden narratives and amplifying marginalized voices.

Sara's humanitarian work shapes everything she writes. As an American Red Cross volunteer, she serves as an International Humanitarian Law educator and youth lead, teaching young people about the Geneva Conventions and the rules that protect the wounded, the prisoners, and the dead. Through Service to Armed Forces, she visits veterans in VA hospitals monthly, witnessing the long aftermath of war. Through Restoring Family Links, she helps refugees and asylum-seeking women search for family members separated by conflict and displacement—continuing work that Clara Barton began over 150 years ago. She also volunteers with the Women of the World refugee center, combining her humanitarian commitments to serve forcibly displaced women.

This book emerged from Sara's own inheritance. Her great-grandfather survived World War I but came home blind; his memoirs describe life among the dead in the trenches. Her grandfather flew gliders at nineteen in World War II, was held as a POW, and only spoke of gathering his comrades' broken bodies in the final days of his life. Her uncles served in Vietnam and never spoke of it at all.

In 2023, Sara stood in a field in Foy, outside Bastogne, Belgium, where her husband's great-uncle had been buried—then disinterred and shipped home. In that moment, she saw the invisible young men who had done that work, and she knew their story had never been told.

Known Unto Us is her testimony to the witnesses who came before.

Sara is currently pursuing graduate studies in archaeology and heritage. She lives in Utah with her spouse and four dogs.